Fine Homebuilding on More Frame Carpentry

Fine Homebuilding® *on*

More Frame Carpentry

The Taunton Press

Library of Congress Cataloging-in-Publication Data

More frame carpentry.
Fine homebuilding on more frame carpentry.
p. cm.
Contains 28 articles from recent issues of Fine homebuilding magazine.
Published simultaneously in hardcover as: More frame carpentry.
"A Fine Homebuilding book" — T.p. verso.
Includes index.
ISBN 1-56158-050-3 (pbk.)
1. House framing — Miscellanea. I. Fine Homebuilding. II. Title
TH2301.M58 1993b 93-9977
694'.2 — dc20 CIP

Cover photo: Scott McBride

First printing: June 1993

Printed in the United States of America.

A Fine Homebuilding Book

The Taunton Press, Inc.
63 South Main Street
Box 5506
Newtown, Connecticut
06470-5506

CONTENTS

INTRODUCTION

You can't know too much about framing a house or an addition. The more you know, the more options you have open to you. And a detailed understanding of the work is sure to save you time and money and improve your results.

That's why we've come out with *More Frame Carpentry.* This volume brings together 28 articles from recent issues of *Fine Homebuilding* magazine –all published since our first volume on frame carpentry came out in 1990. As then, the authors here are experienced builders, writing about the tools, materials and techniques they know best. So whether you're framing a gable roof or installing a garage door, working with treated lumber or with glulams, this volume will help you do your best work.

Jon Miller, editor

Note: A footnote with each article tells you when it was originally published. Product availability, suppliers' addresses and prices may have changed since the article first appeared.

Framing a Second-Story Addition

A quartet of gables is linked by California-style valleys

by Alexander Brennen

Almost every day the local refuse company parks a debris bin in front of some house in Albany, California. The bin usually means that a crowbar-wielding crew will soon arrive to tear the roof off the house, signaling the start of another second-story addition.

In this neighborhood of two-bedroom, wartime tract houses it's commonplace for families to add another level of living space atop their original house. Unfortunately, these additions often show little regard for the original shape of the building or the shadows cast across the neighbor's house and yard to the north. Recently, my partner Michael Keenan and I added a second story to a house in Albany that we think is in keeping with the style of the original structure. Our clients, Joe and Denise Lahr, felt that one of the reasons many second-story additions look out of place is that their height is out of proportion to the original house. To make sure their addition remained in scale with the rest of the building, the Lahrs wanted to keep the highest portion of the roof well back from the sidewalk, with the roofs stepping up as they moved toward the back of the house. The footprint of the original building was basically a rectangle, but it had an appealing roofline because its relatively steep 9-in-12 gable roofs intersected one another in a pleasing asymmetrical geometry of differing ridge heights.

The Lahrs' architect, H. M. Wu, met their needs with a second floor topped with an arrangement of four gables organized around a central hip roof. In plan, the configuration looks a little like a pinwheel (roof plan, facing page). Each gable shelters a distinct part of the addition—one for the sitting room, one for the bathroom, and one each for the two bedrooms. In the center of the plan, a skylit stairwell leads to the upstairs hallway (floor plan, facing page).

The architect kept the walls as low as possible by giving them a 6-ft. 6-in. plate height. Except for the hallway, each room has a cathedral ceiling, so the low plate doesn't make the rooms feel claustrophobic. The flat ceiling over the hallway is 7 ft. 6 in. high, which leaves enough room above it to run heater ducts for the upstairs rooms.

Hip rafters extending upward from the ridges of each gable meet at the main ridge, forming the tallest of the new roofs. This big hip roof links the four new gables, and repeats the form of the original hipped portion of the roof (photo above).

First, remove the roof—We started our work by stripping the shingles from the south side

From *Fine Homebuilding* magazine (February 1990) 58:76-79

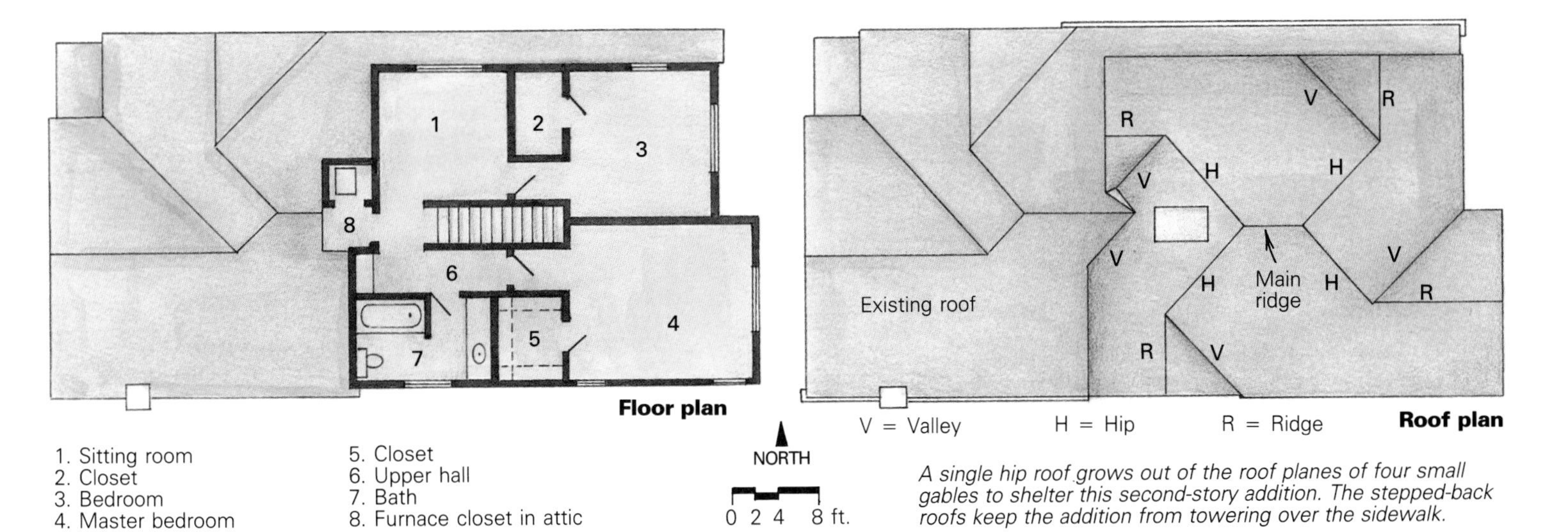

A single hip roof grows out of the roof planes of four small gables to shelter this second-story addition. The stepped-back roofs keep the addition from towering over the sidewalk.

of the old roof, letting them fall onto the driveway. Then we pulled out the 1x8 sheathing and the 2x4 rafters. The north side of the roof was very close to the property line, and we didn't want the neighbor's yard cluttered with roof debris. To prevent that, we cut the north side of the old roof into 3-ft. sq. chunks, using a worm-drive circular saw and a 24-tooth carbide blade with wide-set teeth. The wide-set teeth keep the blade from building up a thick layer of asphalt while cutting through the shingles, and the carbide stands up to the many nails that get cut during this kind of work. Since completing this job, we've started using the "negative-rake" carbide blades for this kind of demolition work.

New piers, new beam—The original footings of the house were sturdy enough to take the weight of the addition, but calculations revealed that raising the original wall would shade the neighbor's solar panels. To keep the sun shining on the panels, the Lahrs asked that the addition's north wall be placed 3 ft. south of the existing foundation line. For solid bearing, we needed a beam supported by two 3-ft. sq. by 18-in. deep piers under the house. We used army shovels to excavate the pier holes in the cramped crawl space.

Excavating footings in a crawl space is just plain disagreeable. Fortunately, it's pretty easy to get the concrete under the house with the variety of pumps available today. We used a grout pump with a 2-in. hose for this pour. The small-diameter hose is really maneuverable in a confined space, but it requires small aggregate—⅜-in. pea gravel—and six sacks of cement per yard of concrete (instead of the typical five) to reach a strength of 2,500 psi.

Piers in place, we turned our attention to the original south wall of the house. The second-floor joists of the new addition would bear on the wall's top plate, so we wanted to make sure that it was level. If it wasn't, subsequent construction would require tedious adjustments. Most of the houses around here have settled somewhat, and this one was no exception; we had to remedy the situation.

The existing ceiling joists (over which we were framing the new floor) were rough-sawn 2x4s varying in width from 3½ in. to 4 in. Between each ceiling joist we nailed a pair of 2x4 blocks, face down, using four 16d box nails in each block. The doubled blocks added 3⅛ in. to the height of the top plate. We then notched each ceiling joist to be slightly below the doubled blocks, and ran a continuous plate across all the blocks and joists as a base for the second-story floor joists.

Next we set up the transit on top of the old ceiling joists to check the plate for level. As an aside, I should mention the importance of setting up the transit over intersecting walls so that the tripod legs have solid bearing. If you set up mid-span over flexible joists, you may spend time wondering just when your transit went out of adjustment.

We checked the heights of all the surfaces we were going to build upon, and sure enough, the building was tilted. We were, however, surprised at the severity of the problem. One end of the wall was 2½ in. lower than the other end. We ripped 2x4s into long shims with a bandsaw to level things up. This is the kind of discrepancy that stucco is very good at hiding.

We positioned our new footings in the crawl space so that they fell directly below existing walls of the downstairs bedroom closets. That allowed us to insert the new posts into the walls from the closet sides of the walls, thereby preserving the finished wall surface in the adjoining bedroom and bathroom. The post on the east side of the house tucks into the exterior wall, and we installed it from the outside.

The beam that rides on the three posts is built up from three 2x12s, spiked every 8 in. with 20d nails. Because the beam is wider than the wall that it rests on it, we aligned its inboard edge with the inside of the new wall. The 2x8 joists of the new floor are supported by joist hangers and a ledger nailed to the side of the beam (drawing next page).

Framing the gables—Layout of the stairwell and joists was next, and to nobody's surprise the existing building was not square. After nailing down our plywood subfloor, we squared the new outside walls as best we could by dividing the error as we snapped our chalklines for the wall. At one end the wall plates project 3/16 in. beyond the edge of the subfloor, while they fall 3/16 in. inside the edge of the subfloor at the other end. Once again, the stucco would hide the discrepancies.

Before framing the rake walls at the end of each little gable, we snapped chalklines on the subfloor to mark their full scale dimensions, along with all stud and header locations. That allowed us to take direct measurements for the angled studs that intersect the top plates. We positioned the chalkline for the bottom plate to coincide with the plate's baseline so that we could frame the first wall, stand it up and nail it in place, and then use the same layout marks to frame the opposite wall.

In order to keep the sidewalls low, we had to put the window headers on top of the wall plates and affix the rafters to them with joist hangers. The headers are 4x6s with a bevel along one edge to match the 9-in-12 roof pitch (drawing next page). Once the exterior walls were sheathed with ½-in. plywood, we placed our doubled 2x10 ridge beams and began installing the common rafters (bottom left photo, p. 11).

Three pairs of common rafters meet at the main ridge. To set this ridge, we first cut the common rafters that join it, and adjusted their length to accommodate the thickness of the ridge. Then we nailed the tails of the rafters to the wall plates, while aligning the ridge by hand between them. In this manner the height of the ridge was automatically set, and we could then bring a bearing wall from below to carry its load.

As anyone who has done any framing knows, it is especially gratifying to see the walls and rafters in place. All the drudge work in the crawl space starts to pay off when the bones of the building begin to give it shape. On this job, everything had been going well—too well.

After setting the main ridge and some of its common rafters, we noticed that the inboard gable-end wall in the master bedroom had gone out of plumb where it met the exterior wall. When building this rake wall we had been unable to install a post directly under

Drawings: Michael Mandarano

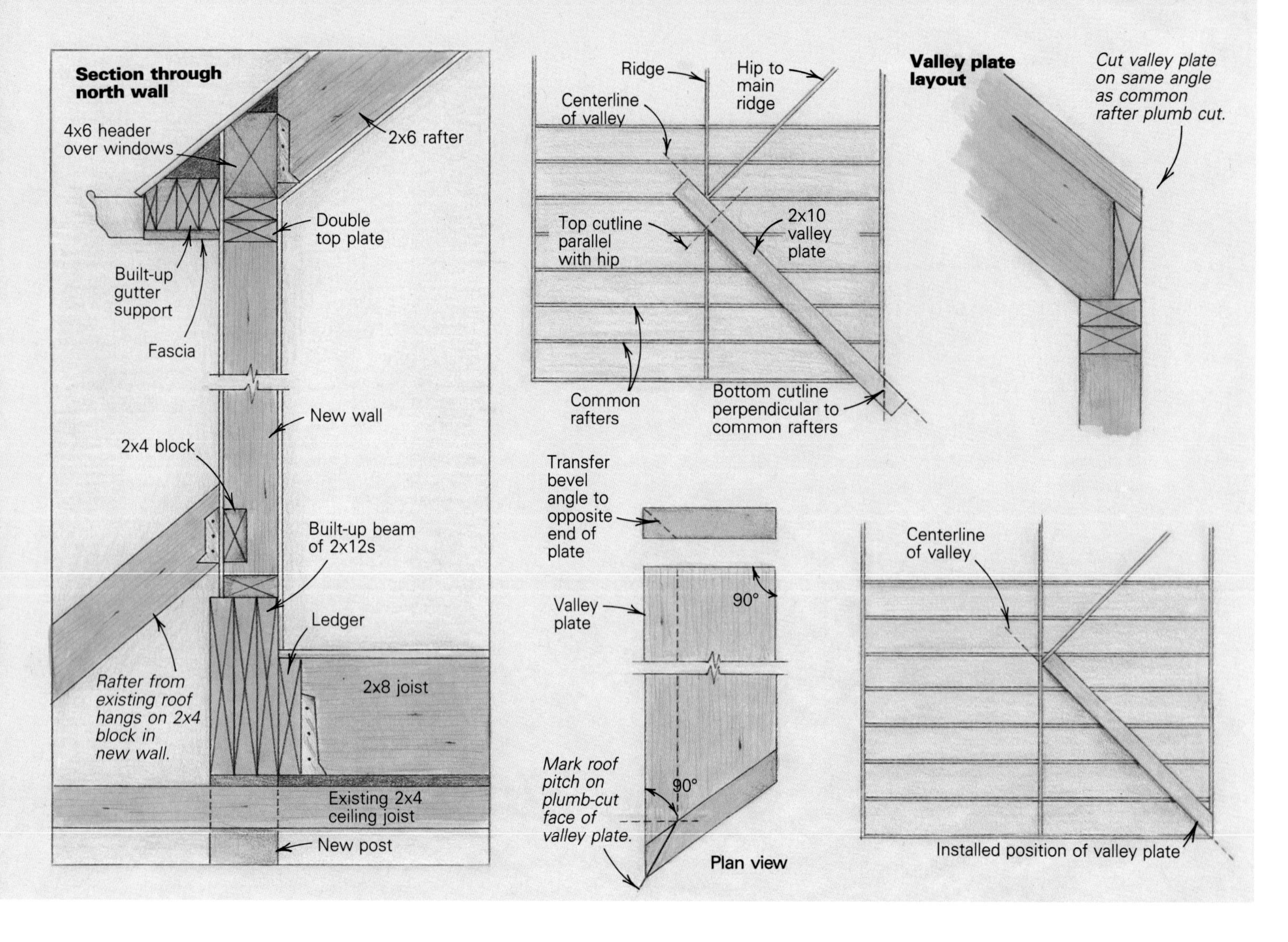

its ridge because of plans to install a heating duct there. There were no horizontal top plates to stop the top of the wall from spreading as we added the weight of the rafters in that bedroom, as well as the rafters supporting the main ridge. When we spotted the problem the wall was out of plumb by ¾ in., and we had the uneasy feeling that the error was compounding at a slow but steady pace. We needed a quick and effective fix.

The interior east/west wall of the master bedroom was still plumb, so we braced it with angled studs nailed to the floor and tied a come-along to its top plate. Then we looped the come-along cable around the top plate of the exterior wall. Before winching the wall back toward the house, we placed floor jacks underneath the main ridge of the house and the ridge of the master bedroom. That allowed us to relieve the load on the exterior wall that was driving it outward while we winched the wall back a little past plumb. Then we sheathed the overloaded wall with ½-in. CDX plywood and linked the opposing rafters with 2x4 collar ties a couple of feet down from the ridge. The wall did spring back to about ⅛ in. out of plumb after we removed the jacks and the come-along. That's a discrepancy we can live with.

California framing—With the main ridge set and the frame bolstered against movement, we were ready to install the four hips and four valleys. As shown in the roof plan (previous page), the hips extend from the ridge beam of each gable to the main ridge beam. Because the hips meet their ridge at 45° in plan, we made the double cheek plumb cuts on their ends with the circular saw set at 45° (for more on the mechanics and theory of roof framing, see pp. 84-89).

Our crew uses the "California-style" valley for roofs. Instead of jack rafters from intersecting ridges meeting at a valley rafter, a California valley is built on a 2x plate that lies flat atop the common rafters of one of the gables (bottom right and top photos, facing page). We use this method because it takes less time to build than conventional valleys, and in this case, we wanted as many common rafters as possible in the individual gables because the drywall ceiling is affixed directly to their bottom edges.

To make a California-style valley, we first snapped a line across the tops of the common rafters to mark the centerline of the valley where the two roofs would intersect. Then we laid a 2x10 next to the chalkline, as shown in the drawing above. Using a straightedge, we marked two cuts on the plate. The bottom cut is perpendicular to the common rafters, while the top cut is parallel with the hip rafter. The length is found by measuring in place.

We made the bottom cut with the circular saw set to the plumb cut for a 9-in-12 roof, which works out to 37°. This puts the face of the cut in plane with the sides of the intersecting rafters. On the face of the cut, we marked a 9-in-12 pitch line to represent the plane of the tops of the interesecting rafters. The leading edge of the valley plate should be beveled to this line. The plan view of the plate shows how the pitch line can be transferred to the opposite end of the plate, where you can make a direct degree reading. We typically bevel these plates on our 12-in. bandsaw before nailing them in place to the common rafters. Then we install blocks next to the plate, between each common rafter, as a nailing surface for the roof plywood.

We measured the jack rafters between the valley and the ridge in place. Their tops had the plumb cut angle of the common rafters, but with a 45° cheek cut to meet the hip rafters. Their bottom cuts were level, with the saw set at 37° to match the 9-in-12 pitch of the valley plate.

Before we sheathed the roof we laminated three 2x4s and a 1x4 directly to the walls for a gutter support that doubles as a narrow eave (section drawing above). We beveled the tops of the two outer laminations to be in line with the roof plane and finished the detail with a fascia board affixed to the bottom of the gutter support. □

Alexander Brennen is a partner in Zanderbuilt Construction in Berkeley, California.

Framing details. **Before the hip roof was erected, Brennen and his crew framed the four small gables that enclose the upstairs rooms (photo above). The joists in the foreground will carry the ceiling of the hallway. To the left is the skylight well. A California-framed valley relies on a valley plate for jack-rafter bearing rather than on a valley rafter (photo right). The plate is nailed to the common rafters, and its leading edge is aligned with the centerline of the valley. In the photo at the top of the page, the author aligns the lower end of the plate with the valley's centerline, while Michael Keenan snugs the upper end of the plate against a hip rafter leading to the main ridge.**

Permanent Wood Foundations

A cost-effective and energy-efficient alternative to concrete

by Bill Eich

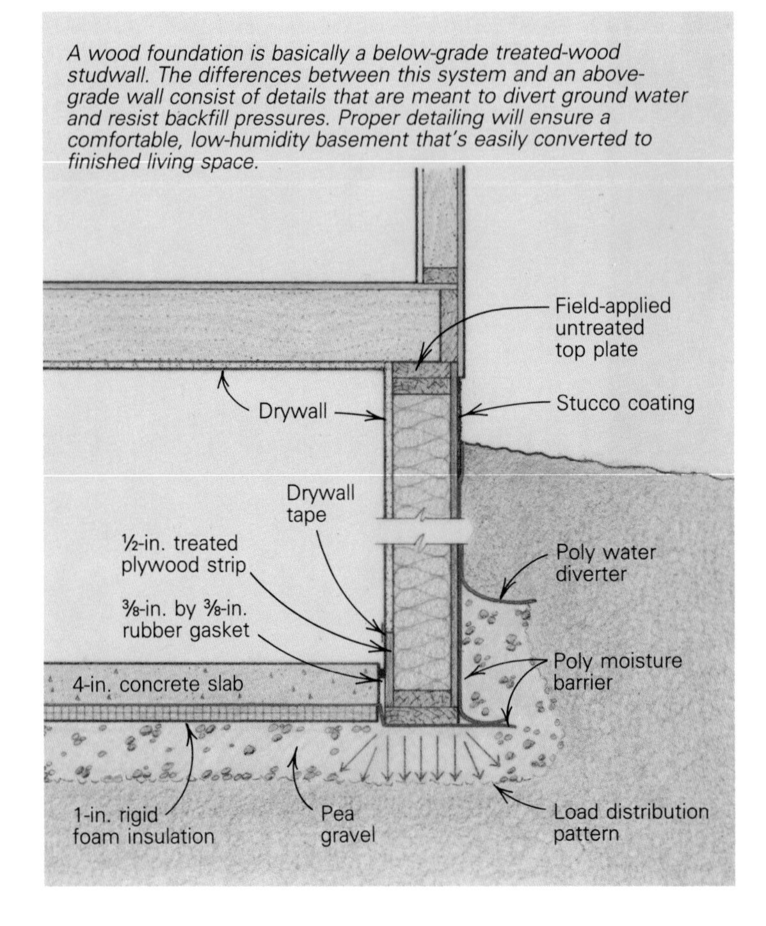

A wood foundation is basically a below-grade treated-wood studwall. The differences between this system and an above-grade wall consist of details that are meant to divert ground water and resist backfill pressures. Proper detailing will ensure a comfortable, low-humidity basement that's easily converted to finished living space.

The permanent wood foundation (PWF) is an engineered foundation system that's made with perservative-treated wood, and my company has been installing them for more than ten years. Though we were initially skeptical, our experiences have made us believers in the system. As a company that specializes in the design and construction of energy-efficient, airtight housing, we require basements that are warm, dry and structurally sound. The PWF has all of these qualities, and at a cost that's competitive in this area with poured concrete and concrete block.

PWFs have been available for many years (they were originally known as All Weather Wood Foundations; see *FHB* #5, pp. 40-42), yet they account for a mere 5% of the U. S. market and about 20% of the Canadian market. Buyer resistance isn't the problem, though; When both systems are explained and the benefits compared, 75% of our clients in the $100,000 to $200,000 price range choose wood over concrete. The main obstacles to more widespread use of the system have been builder resistance and a powerful concrete-industry lobby that has slowed code approval in many jurisdictions. In spite of these obstacles, however, PWFs are now approved by all major codes in America and Canada, though a few cities still prohibit or limit their use. In this article I'll describe why I think the PWF is such a good system. I'll also give an overview of how to install one, including some tricks we've learned over the past decade that aren't included in the official installation guides.

Low cost, high comfort—From the consumer's viewpoint, the benefits of a wood foundation are warmth, dryness and finishability. A wood-foundation basement can be easily converted to an inexpensive living space that's as comfortable as the main level. In areas where residential construction costs average $60 to $70 per sq. ft., a wood basement can normally be converted to finished space for $10 to $20 per sq. ft. Even in an unfinished basement, R-19 insulation batts installed from floor to ceiling will reduce both heat loss and utility bills. And the superior drainage features of the PWF ensure a dry basement storage area that's free from mold and mildew.

From a builder's perspective, the benefits of a PWF are cost and control. It costs us $20 to $25 per lineal foot to install a wood foundation, while our subcontractors charge us $30 to $35 for a masonry wall. Using our own framing crew to install foundations gives us much more scheduling flexibility than if we had to depend on a masonry subcontractor. And a PWF can be installed during the winter without expensive shelters or temporary heating units. At most, a little preplanning is required. Spreading a layer of straw or foam insulation over the building site keeps the ground from freezing too deeply for excavation. This lets us start jobs with wood foundations in January or February that we would otherwise postpone until April.

Two of the first five PWFs in America were built and installed in 1971 here in Spirit Lake, Iowa, by Citation Homes, one of the nation's leading suppliers of factory-fabricated wood-foundation panels. Those foundations are performing today, more than 20 years later, as efficiently as the day they were installed. And though no one knows what the ultimate lifespan of a typical foundation will be, at least one company guarantees them for 75 years.

Sticks and stones—A PWF is a load-bearing, lumber-framed wall that's sheathed with plywood. It sits on a concrete footing or a bed of gravel or crushed stone (drawing left). A wood foundation can be framed with 2x6s or 2x8s, depending on the particular loading requirements. It goes together much like a standard studwall, with studs, plates and plywood. But there are important differences. The need to resist backfill pressure creates more critical loading and stress requirements than are present above grade. Because of this, it's important to use the proper fastening and blocking techniques. In fact, most problems with wood foundations can be traced to improper installation by inexperienced workers.

Wood-foundation problems are less expensive to correct than those involving masonry foundations. Even so, builders just learning the system should start by using prefabricated foundation panels that have been properly engineered at a component manufacturing facility. Only after a great deal of experience with factory-fabricated foundation panels should one consider site-building a PWF—it's too easy to miss a small detail and have it come back to haunt you. Though there's no central source of information about panel manufacturers, two good companies I'm aware of are Citation Homes (1100 Lake St., Spirit Lake, Iowa 51360; 712-336-2156) and Permanent Wood Foundations, Inc. (P. O. Box 819, Flint, Mich. 48501; 313-232-5099). The former will ship panels

Drawings: Vince Babak

anywhere from Colorado to Ohio, the latter ships nationwide.

The foundation-grade lumber used in a PWF is pressure-treated with a chromated copper arsenate (CCA) solution to a retention level of .60 lb. of chemical per cu. ft. of wood—50% more preservative than the codes require for ground-contact lumber. Although the chemical is water soluble during treating, it permanently fuses to the wood cells as it dries, making it clean and safe to use. Southern yellow pine allows the highest level of preservative penetration, making it the material of choice for most treated wood foundations (for more on preservative-treated wood, see the article on pp. 114-118).

Laying the footings—Regardless of whether you use site-built or prefabricated panels, the first step is to dig the basement and footings. The basement is excavated 10 in. to 11 in. below the finished basement-floor level. To minimize backfill pressure on the walls, we try to limit the overdig around the perimeter (the extra width and length needed to facilitate working around the outside of the foundation) to between 12 in. and 18 in., particularly where we have a full 7 feet of backfill height. We dig our foundations with an end loader, which often means excavating a ramp into the hole. We try to place this ramp at the lowest point of the final grade around the foundation to further reduce backfill pressure.

After the excavation is complete, we set the perimeter stringlines. Then 12-in. long, treated 2x2 grade stakes are driven every 4 ft. to 6 ft. around the foundation perimeter, just inside the stringlines. We use a transit to set the tops of these stakes 4½ in. below the finish floor level, which leaves them about 6 inches above the basement subgrade. The entire subgrade is then covered with 5 inches to 7 inches of washed pea gravel, which is spread with a skid loader (such as a Bobcat). The perimeter needs to be very level, so it's floated with a 2x4, using the grade stakes as a leveling guide (top photo). Drain tile isn't required with a wood foundation: the pea-gravel base acts as a huge drain tile. We usually install a sump basin in a utility room; any ground water flows into the basin and is pumped away.

Next, we lay the treated footing plate around the perimeter. For 2x4 foundation walls, a 2x6 footing plate is required, while 2x6 walls need a 2x8 plate. The footing plate is nailed down into the 2x2 grade stakes (middle photo). The grade stakes serve no structural function; they are merely a convenience to level the pea gravel and hold the footing plates during the erection process. We've tried several other techniques but the 2x2s work best.

Many builders get nervous about the prospect of letting all foundation loads bear directly on the pea gravel, with no concrete footing. To understand why the technique works, remember that when confined in an enclosed space—such as a basement ditch—pea gravel is non-compressible. You can prove it to yourself by taking a 5-gal. bucket, filling it with pea gravel and striking off the top. Regardless of

Moisture control is crucial to a properly functioning wood foundation (photo above). First, the subgrade is covered with pea gravel, and the perimeter is leveled using grade stakes as a guide.

A treated footing plate is then installed (photo above) over a 2-ft. wide poly strip and is nailed down into the grade stakes.

This strip is tape-sealed (photo below) to a poly moisture barrier installed over 1 in. of rigid-foam insulation.

From *Fine Homebuilding* magazine (June 1991) 48:62-66

how much it's shaken or tamped, it won't settle or compress. The pea gravel spreads the foundation wall load from the footing plate to the subgrade below at a 45° angle. The bearing capacity is thus a function of the depth of the pea gravel: an 8-in. wide footing plate on 6 inches of pea gravel will spread the load over a 20-in. wide area, just like a 20-in. by 8-in. concrete footing.

A 2-ft. wide strip of 4-mil poly is installed beneath the footing plate. It serves as a capillary moisture break. This strip is sealed with housewrap tape to a continuous poly moisture barrier under the entire basement floor (bottom photo, previous page). An ordinary 6-mil poly meets the PWF specification guidelines, but we prefer to use a higher quality 4-mil product called Dura-Tuff (Yunker Industries, 200 Sheridan Springs Rd., Lake Geneva, Wisc. 53147; 414-248-6232), which is both puncture resistant and UV stabilized. If Dura-Tuff is unavailable, we use a product called Rufco (Raven industries, P. O. Box 1007, Sioux Falls, S. D. 57117; 605-336-2750) that has similar properties. We also place 1 in. of rigid foam under all of our basement floors, laying the foam directly over the pea gravel and beneath the moisture barrier. The foam isn't required by the PWF specifications, but using it is good construction practice for comfort, moisture control and energy efficiency. If you don't use foam, you'll have to screed the pea gravel to the top of the footing plate or pour the basement floor 4½ in. thick.

Slab first—At this point, many builders fabricate and erect the foundation wall panels, leaving the basement floor for later. We choose to pour our basement floors first because concrete is a much better work surface than pea gravel. We place our floor forms directly on the footing plate. The edge of the slab is held back ½ in. from where the inside edge of the wall will be. This leaves room for a ½-in. treated plywood strip (bottom photo, right). We used to install this so that the top edge was flush with the concrete floor; now we let it protrude about 2 in. This lets us tape the plywood to the finish drywall, completing the wall air barrier. Sure-Seal, a ⅜-in. by ⅜-in. saturated urethane foam gasket (Denarco Sales, 12710 Idlewild, White Pigeon, Mich. 49099; 616-641-2206) completes the air barrier between the wall and the slab, as well as reducing radon-gas penetration. The concrete floor must extend 1½ in. above the bottom of the studs—it keeps the bottom of the foundation wall and studs in place during the backfill.

Framing the walls—Once the basement floor is finished we begin framing the walls. All fasteners must be corrosion resistant. PWF specifications require double hot-dipped galvanized nails above grade, stainless-steel below. For power nailing, we use type 304 or 326 stainless-steel nails or staples. Senco Products, Inc. (8491 Broadwell Rd., Cincinnati, Ohio 45244; 800-543-4596) makes stainless-steel fasteners that meet the specification for air nailers.

There's a fair amount more crown and twist in treated lumber than in standard framing stock. To compensate for this, we figure on a little more waste and make sure that all crowns face toward the exterior. Doing so sets the studs in opposition to backfill pressure which, in turn, helps straighten the studs. We save the straightest pieces for the corners and plates. Studs are placed with the cut ends at the top of the wall so that we won't have to re-treat the occasional piece that hasn't been fully penetrated by preservative. The wall is sheathed with ½-in. treated plywood, and all joints are sealed with butyl caulk.

Stud size and spacing vary with material grade and backfill depth. In general, though, 42 inches of backfill will require a 2x4 wall framed at 12 in. o. c., 64 inches of backfill a 2x6 wall 16 in. o. c., and 84 inches of backfill a 2x6 wall 12 in. o. c. It's possible to engineer a wood-foundation wall with a full 96 inches of backfill, but the added cost of the required 2x8 framing makes the finished foundation more expensive than concrete.

Window openings require one extra king stud for each stud that's been cut. The rough sill piece on windows must be doubled for window openings up to 6 feet wide and tripled for openings wider than 6 feet. The sill distrib-

Energy detailing. **Eich lets a ½-in. treated plywood strip protrude 2 in. above the basement floor (photo above). It is later taped to the drywall. This, along with a rubber gasket, completes the foundation air barrier. Before the wall is nailed in place (top photo), a small hydraulic jack compresses the rubber gasket between the wall and the basement floor slab to ensure a tight seal at the joint.**

utes the backfill pressure from the front of the window to the king studs at the sides.

After the walls have been framed and sheathed, we fasten the self-adhesive Sure-Seal to the side of the slab and then stand the walls upright. Here we see another advantage of pouring the slab first. Before nailing the walls to the footing plate, we push them against the slab with a small hydraulic jack (top photo, facing page). This not only straightens the bottom plate but it compresses the rubber gasket, ensuring a tight seal between the wall and the slab. A second, untreated top plate is also nailed on at this point.

The first-floor deck needs to absorb and distribute any backfill loads. Because of this, the foundation can't be backfilled until the floor is complete. Care must be taken to toenail each floor joist to the top plate with three 10d nails. The rim joist is also toenailed at least every 12 in. On walls where the floor joists run parallel to the foundation, 2x10 blocking must be installed—24 in. o. c. maximum in the first joist bay and 48 in. o. c. in the second (photo below right). The blocks are secured to the foundation wall with metal framing anchors. The subfloor is glued and nailed to the blocks as well as to the joists. This is an area where proper installation is critical: a poor job of fastening the floor joists or the blocking will cause the wall to fail.

Special considerations—Some details that require no special thought or care with masonry foundations are quite different with wood. A basement stairway that runs along an exterior foundation wall is a good example. Because the stair opening will prevent the floor deck from absorbing the wall loads at that point, a beam must be built to transfer the horizontal loads around the opening to the floor. We build this beam by gluing and bolting six 2x6s flat on top of the foundation wall along the stairway. The beam extends about 2 feet beyond each end of the stairway (drawing right).

Uneven backfill heights are another potential problem, particularly in homes with an 84-in. backfill height at the front of the house and a walkout basement at the rear. In these cases, interior shear-wall panels spaced 15 feet to 20 feet apart will help the foundation resist any racking forces (top photo, next page). A shear wall is a 4-ft. section of wall perpendicular to the foundation wall that's sheathed with ½-in. plywood or oriented strand board and nailed 4 in. o. c. Shear walls are lag-bolted up into the floor joists and down into the basement floor. We try to place these walls where a future interior wall may be located. An alternate, but more expensive technique, is to bury a concrete deadman outside of the foundation wall, tying it to the studs with re-rod stirrups. Finally, whenever a garage floor, sidewalk or patio is poured next to a wood foundation, tying the slab to the foundation with lag bolts or re-rod stirrups will stiffen the top of the foundation (bottom photo, next page). Whatever the permanent support system, it's a good idea to add a few extra temporary braces to the inside of the foundation before backfilling.

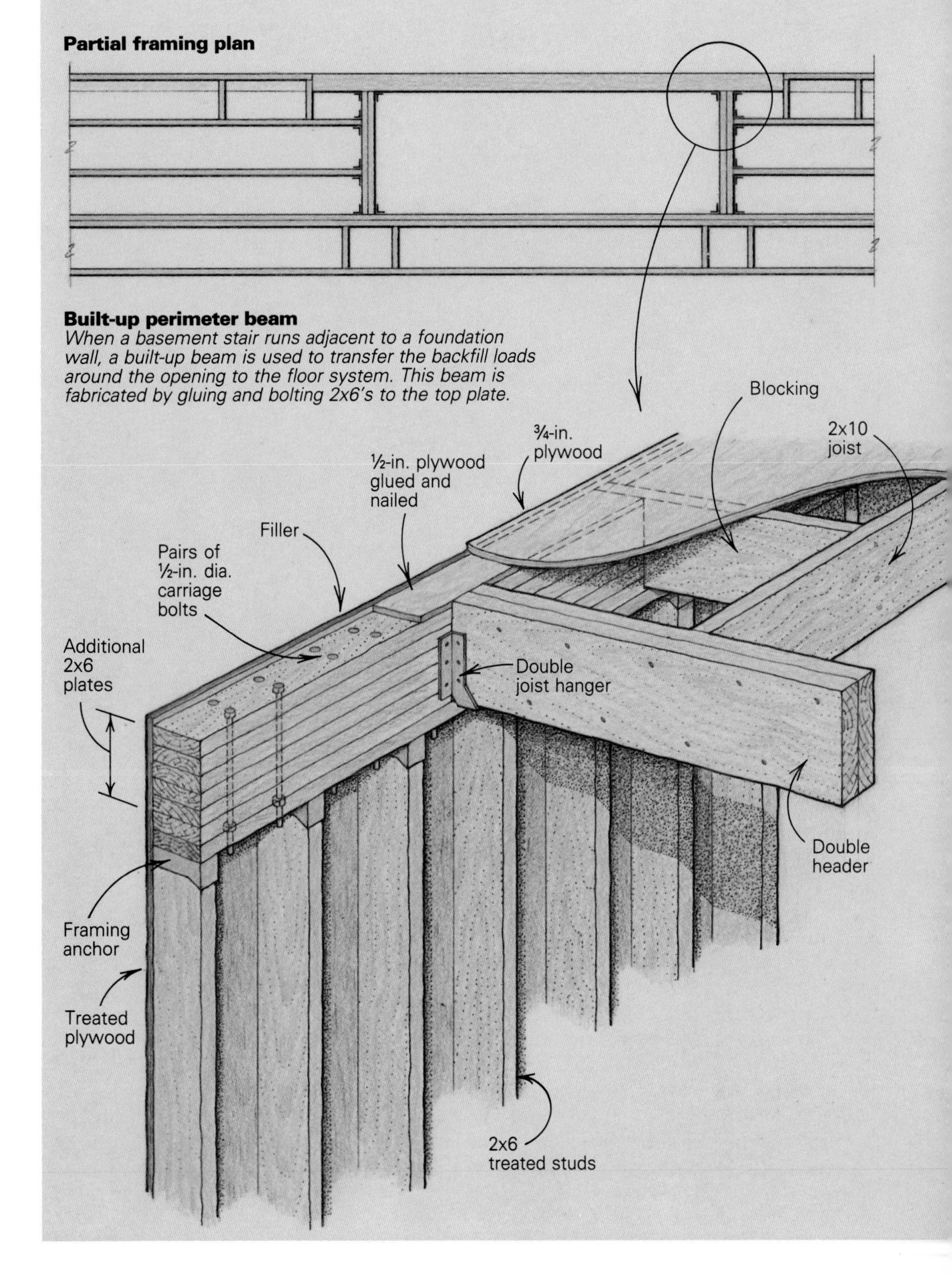

After the walls have been raised and braced, the foundation is wrapped with an 8-ft. wide sheet of poly. We prefer Dura-Tuff for this application, for the same reasons we use it under the slab. The poly acts as a water diverter, an area of low friction that will direct ground water to the pea-gravel footing. The first 12 in. around the foundation is backfilled with pea gravel and topped with a poly strip. We backfill with a skid loader and make sure that the walls are braced. The bracing prevents large rocks and clods of soil from bouncing off the walls and causing them to bow inward.

The above-grade portion of the treated plywood can be finished in a variety of ways. Although 1x12 cedar, ¼-in. cement board and fiber reinforced plastic are all commonly used

To absorb backfill loads, blocking is needed near foundation walls as the underside of this first floor shows. The subflooring is then glue-nailed to the joists and the blocking.

treatments, our first choice is a brush-on stucco coating called Retro Flex (Retro Technologies, Inc., 328 Raemisch Rd., Waunakee, Wisc. 53597; 608-849-9000). It's easily applied and gives the walls the appearance of concrete.

The bowed and twisted pieces of treated wood discarded earlier can be used for bearing-ledge panels or garage foundation panels. They get buried in the ground and don't have to be straight. A bearing-ledge panel is simply an open 2x4 stud panel with studs 16 in. o. c. and a single top and bottom plate (photo above). It's attached to the outside of the basement wall after the poly water diverter is on. Its purpose is to carry brick veneer, a sidewalk, masonry steps or a garage floor.

Garage foundations or slab-on-grade perimeter foundations can also be built with treated wood. These typically consist of 2x4 studs, 36 in. to 48 in. long, with a double top plate and a single 2x6 sill plate that also serves as a footing plate. To provide some shear strength, the top 24 in. of the foundation panels are sheathed with ½-in. plywood. Garage foundations usually protrude 6 in. to 12 in. above the floor, so 12 in. of plywood is also nailed to the inside to allow a cleaner concrete job. For a slab-on-grade system, the concrete floor is poured flush with the top of the wall and the inside plywood is eliminated. Whether garage or slab-on-grade, we brace these walls thoroughly before backfilling—there's no slab to hold the bottom in place and no floor framing to hold the top. Even with the extra bracing, we must still sometimes re-align the garage panels with the skid loader after they've been backfilled.

Where backfill heights vary widely, interior shear walls resist backfill pressure (top photo). Tying the foundation to masonry adds even more stability. The box that's being framed outside the foundation is a bearing-ledge panel. It will eventually carry a concrete stoop. Re-rod stirrups (photo above) will secure the foundation to an adjacent driveway or slab.

Early experiences—A final, rarely mentioned advantage of wood foundations is that they're more forgiving of error than are concrete ones. One foundation we installed was backfilled during the winter with a combination of snow and large dirt clods. When the spring thaw came, a 2-ft. deep trench appeared all around the house, followed immediately by 2 in. of rain. As I approached the house after the rain stopped, I could see the water line where the basement ditch virtually formed a moat. Expecting the worst, I found instead a perfectly dry basement and a hard-working sump pump.

Another wood foundation we installed was quickly capped off with a floor system early one winter in a snow storm. We didn't return to frame the house until spring, when we noticed that one of the end walls had bowed in about 2 ft. at the top. After closer examination, it was obvious that in their haste to get out of the snow, the crew had forgotten to install the metal framing anchors on the endwall blocks. I brought in the backhoe, dug out the backfill, pushed the wall back out with one hand, nailed the framing anchors in place properly and backfilled again. The problem was fixed in one hour at a cost of $45. A failing concrete wall would have been much more expensive to fix. □

Bill Eich is a custom-home builder from Spirit Lake, Iowa, and is president of the Energy-Efficient Building Association. Photos by Charles Wardell. For additional technical information, Permanent Wood Foundation: Guide To Design and Construction, *58 pp. is available from the American Plywood Association (P. O. Box 11700, Tacoma, Washington 98411; 206-565-6600).*

Framing a Gable Roof

Cutting and stacking the way pieceworkers do it

by Larry Haun

The key to production roof framing is to minimize wasted motion. Here, carpenters nail gable studs plumb by eye—there's no need to lay out the top plates. The next step is to snap a line across the rafter tails and cut the tails off with a circular saw.

One of my earliest and fondest memories dates from the 1930s. I remember watching a carpenter laying out rafters, cutting them with a handsaw, and then over the next several days, artfully and precisely constructing a gable roof. His work had a fascinating, almost Zen-like quality to it. In a hundred imperceptible ways, the roof became an extension of the man.

But times change, and the roof that took that carpenter days to build now takes pieceworkers (craftspersons who get paid by the piece and not by the hour) a matter of hours. Since they first appeared on job sites, pieceworkers have given us new tools, ingenious new methods of construction and many efficient shortcuts. But what skilled pieceworkers haven't done is sacrifice sound construction principles for the sake of increased production. Quite the opposite is true; they've developed solid construction procedures that allow them to keep up with demand, yet still construct a well-built home.

The secret to successful piecework, from hanging doors (see *FHB* #53, pp. 38-42) to framing roofs, is to break down a process into a series of simple steps. To demonstrate just how easy roof framing can be (with a little practice), I'll describe how to cut and stack a gable roof the way pieceworkers do it.

The rafter horse—To begin with, pieceworkers try to avoid cutting one piece at a time. They'll build a pair of simple horses out of 2x stock so that they can stack the rafters on edge and mark and cut them all at once. To build the rafter horses, lay four 3-ft. long 2x6s flat and nail a pair of 2x blocks onto each, with a 1½ in. gap between them so that you can slip in a long 2x6 or 2x8 on edge (photo left, following page). An alternate method is to cut a notch 1½ in. wide by about 4 in. deep into four scraps of 4x12. Then you can slip a long 2x6 or 2x8 on edge into these notches. Either of these horses can easily be broken down and carried from job to job. The horses hold the rafters off the ground, providing plenty of clearance for cutting.

Cutting the rafters—Rafters can be cut using a standard 7¼-in. sidewinder or worm-drive circular saw. This isn't the first choice for most pieceworkers, who prefer to use more specialized tools (especially when cutting simple gable roofs). But it is the more affordable choice for most custom-home builders. If you are using a standard circular saw, load rafter stock on the horses with their crowns, or convex edges, facing *up*—same as the rafters will be oriented in the roof frame. Determine which end of the stack will receive the plumb cuts for the ridge and flush this end. An easy way to do this is to hold a stud against the ends and pull all the rafters up against it. Then measure down from this end on the two outside rafters in the stack and make a mark

From *Fine Homebuilding* magazine (April 1990) 60:83-87

Pieceworkers typically build a pair of simple portable rafter horses (photo above). The horses allow them to stack rafters off the ground on edge so that they can mark and cut the rafters all at once.

One way to lay out rafters is with a site-built rafter tee (photo above). The fence on top of the tee allows easy scribing of the ridge cuts and birds' mouths. When laid out this way, rafters are cut one at a time with a circular saw.

corresponding to the heel cut of the bird's mouth (the notch in the rafter that fits over the top plate and consists of a plumb heel cut and a level seat cut). Snap a line across the tops of the rafters to connect the marks.

Next, make a rafter pattern, or layout tee, for scribing the ridge cuts and birds' mouths (right photo above). I usually start with a 2-ft. long 1x the same width as the rafters. Using a triangular square such as a Speed Square (The Swanson Tool Co., 1010 Lambrecht Rd., Frankfort, Il. 60423), scribe the ridge cut at one end of the template. Then move down the template about one foot and scribe the heel cut of the bird's mouth, transferring this line across the top edge of the template. This will serve as your registration mark when laying out the birds' mouths.

The layout of the bird's mouth on the tee depends on the size of the rafters. For 2x4 rafters, which are still used occasionally around here, measure 2½ in. down the plumb line and scribe the seat cut of the bird's mouth perpendicular to the plumb line. Leaving 2½ in. of stock above the plate ensures a strong rafter tail on 2x4 rafter stock. One drawback to this is that for roof pitches greater than 4-in-12, 2x4 rafters will have less than a 3½-in. long seat cut. Consequently, the rafters won't have full bearing on a 2x4 top plate. However, this presents no problems structurally as long as the rafters are stacked, nailed and blocked properly (the building code in Los Angeles requires a minimum bearing of 1½ in.). For 2x6 or larger rafters, you can make the seat cuts 3½ in. long without weakening the tails.

Once the layout tee is marked and cut, nail a 1x2 fence to the upper edge of the tee. This allows you to place the tee on a rafter and mark it quickly and accurately. Make sure you position the fence so that it won't keep you from seeing the ridge cut or your registration mark. Use the layout tee to mark the ridge cut and bird's mouth on each rafter. Scribe all the ridges first at the flush ends of the stock, sliding the rafters over one at a time. Then do the same for the seat cuts, aligning the registration mark on the template with the chalk marks on the rafters.

Next, with the rafters on edge, cut the ridges with your circular saw, again moving the rafters over one at a time. Then flip the rafters on their sides and make the seat cuts, overcutting just enough to remove the birds' mouths.

Production rafter-cutting—Cutting common rafters with production tools is both faster and easier than the method I've just described. In this case, you'll want to stack the rafters on edge, but with their crowns facing *down*. Flush up the rafters on one end and snap a chalkline across them about 3 in. down from the flush ends (the greater the roof pitch and rafter width, the greater this distance). The chalkline corresponds to the short point of the ridge plumb cut. Snap another line the appropriate distance (the common-rafter length) from this point to mark the heel cuts of the birds' mouths. Then measure back up from this mark about 2½ in. and snap a third line to mark the seat cut of the bird's mouth. This measurement will vary depending on the size of the rafters, the pitch of the roof and the cutting capacity of your saw (more on that later).

Now gang-cut the ridge cuts using a beam saw (top right photo, facing page). Blocks nailed to the top of the rafter horses will help hold the stack upright. My 16-in. Makita beam saw will cut through a 2x4 on edge at more than an 8-in-12 pitch and will saw most of the way through a 2x6 at a 4-in-12 pitch. To determine the angle at which to set your saw, use a calculator with a tangent key or, just as easy, look up the angle in your rafter-table book.

For steeper pitches or wider stock, make a single pass with the beam saw (or a standard circular saw) and then finish each cut with a standard circular saw, moving the rafters over one at a time. This way the only mark needed is the chalkline. The kerf from the first cut will accurately guide the second cut.

To make the process go even faster, apply paraffin to the sawblade and shoe. Also, try to stay close to your power source. If you have to roll out 100 ft. of cord or more, the saw will lose some power and won't operate at its maximum efficiency.

Another method for cutting ridges is to use the Linear Link model VCS-12 saw (Progressive Power Tools Corp., 303 N. Rose St., Suite 304, Kalamazoo, Mich. 49007). The model VCS-12 is a Skil worm-drive saw fitted with a bar and cutting chain that lets the saw cut to a depth of 12 in. at 90° (see *FHB* #39, p. 90). It's adjustable to cut angles up to 45° (top left photo, facing page). You can buy the saw or a conversion kit that will fit any Skil worm drive.

With the right tools, the birds' mouths can also be gang-cut with the rafters on edge. For the heel cuts, set your worm-drive saw to the same angle as the ridge cut and to the proper depth, and then make a single cut across all the rafters (bottom left photo, facing page). Seat cuts are made using a 7¼-in. or, better yet, 8¼-in. worm-drive saw fitted with a swing table. A swing table replaces the saw's standard saw base and allows the saw to be tilted to angles up to 68° (bottom right photo, facing page). I bought mine from Pairis Enterprises and Manufacturing (P. O. Box 436, Walnut, Calif. 91789). Set the swing table to 90° minus the plumb-cut angle (for example, 63½° for a 6-in-12 roof) and make the seat cuts, again in one pass.

The only drawback to using a swing table with a worm-drive saw is that it won't allow a substantial depth of cut at sharp angles, so it limits the amount of bearing that the rafters will have on the top plates (about 2½ in. maximum with an 8¼-in. saw). Again, this is of little concern if the roof is framed properly. Nevertheless, for jobs requiring a greater depth of cut, Pairis Enterprises just introduced a swing table to fit 16-in. Makita beam saws.

Gang-cutting birds' mouths works especially well because you needn't overcut the heel or the seat cut, which weakens the tail. Once you get used to working with these production tools, you'll find that it takes longer to stack the rafters than to cut them.

An even faster way to make the seat cuts is to use an 8¼-in. worm-drive saw equipped with a universal dado kit, a rig that has been around the tracts for over 15 years (middle photo, facing page). The dado kit (also manufactured by Pairis Enterprises) consists of an accessory arbor that fits on the saw, allowing it to accept a stack of carbide blades up to 3¼ in. thick. With this setup, birds' mouths can be gang-cut in a single pass and require just one chalkline for the heel cut.

The rig is surprisingly easy to control as long as it's used for its intended purpose, which is to plow out stock on a horizontal surface. In use, the rig whines like a router and hurls big chunks of stock out the front end. Though the guard effectively prevents wood chips from hitting the operator in the face, it's particularly important to wear safety glasses when operating this tool. The only drawback to this dado saw is its cost—about $750 including the saw. But if you cut a number of roofs a year, it will pay for itself in short order.

With the rafters cut, you can now carry them over to the house and lean them against the walls, ridge-end up. The rafter tails will be cut to length in place later.

Staging and layout—Now it's time to prepare a sturdy platform from which to frame the roof. The easiest way is to simply tack 1x6s or strips of plywood across the joists below the ridge line to create a catwalk (the joists are usually installed before the roof framing begins). Run this catwalk the full length of the building. If the ridge works out to be higher than about 6 ft., pieceworkers will usually frame and brace the bare bones of the roof off the catwalk and then install the rest of the rafters while walking the ridge.

For added convenience, most roof stackers install a hook on their worm-drive saws that allows them to hang their saw from a joist or rafter. When not in use, the hook folds back against the saw and out of the way (for more on these saw hooks, see *FHB* #55, p. 92).

The next step is to lay out the ridge. Most codes require the ridge to be one size larger than the rafters to ensure proper bearing (2x4 rafters require a 1x6 or 2x6 ridge). Make sure to use straight stock for the ridge. In the likely event that more than one ridge board is required to run the length of the building, cut the boards to length so that each joint falls in the center of a rafter pair. The rafters will then help to hold the ridge together. Let the last ridge board run long—it will be cut to length after the roof is assembled.

Be sure to align the layout of the ridge to that of the joists so that the rafters and joists will tie together at the plate line. If the rafters and joists are both spaced 16 in. o. c., each rafter will tie into a joist. If the joists are spaced 16 in. o. c. and the rafters 24 in. o. c., then a rafter will tie into every fourth joist. Regardless, no layout is necessary on the top plates for the rafters. Rafters will either fall next to a joist or be spaced the proper distance apart by frieze blocks installed between them. Once the ridge is marked and cut, lay the boards end to end on top of the catwalk.

Nailing it up—Installation of the roof can be accomplished easily by two carpenters. The first step is to pull up a rafter at the gable end.

To save time, ridges can be gang-cut with a 16-in. beam saw (photo right). Though these saws won't cut all the way through anything wider than a 2x4 at a 4-in-12 pitch, where necessary each cut can be completed using a standard circular saw. For these finish cuts, the kerfs guide the saw. The only layout required is a single chalkline across the top edge of the rafters. An alternate method is to use a Linear Link saw (photo left), a Skil worm-drive saw fitted with a bar and chain.

Photo: Eric Haun

One method for gang-cutting birds' mouths is to cut the heels with a worm-drive saw (photo left) and the seats with a worm-drive saw fitted with a swing table, an accessory base that adjusts from -5° to 68° (photo right). By equipping a saw with a universal dado kit (photo above), birds' mouths up to 3¼ in. wide can be plowed out in a single pass.

Top left photo courtesy of Progressive Power Tools Corp.

While one carpenter holds up the rafter at the ridge, the other toenails the bottom end of the rafter to the plate with two 16d nails on one side and one 16d nail (or backnail) on the other. The process is repeated with the opposing rafter. The two rafters meet in the middle and hold each other up temporarily, unless, of course, you're framing in a Wyoming wind. If that's the case, nail a temporary 1x brace diagonally from the rafters to a joist.

Next, move to the opposite end of the first ridge section and toenail another rafter pair in the same way. Now reach down and pull up the ridge between the two rafter pairs. There is no need to predetermine the ridge height (photo facing page). Drive two nails straight through the ridge and into the end of the first rafter, then angle two more through the ridge into the opposing rafter. To keep from dulling a sawblade when you're sheathing the roof, avoid nailing into the top edge of a rafter. At this point, nail a 2x4 leg to the ridge board at both ends to give it extra support. If these legs need to be cut to two different lengths to fit beneath the ridge, it means that the walls probably aren't parallel and, consequently, that the ridge board isn't level. In this case yank the nails out of the rafter pair at the top plate on the high end of the ridge and slide out the rafters until the ridge is level. The key to avoiding all this hassle is, of course, to make sure the walls are framed accurately in the first place.

Next, plumb this ridge section. This can be accomplished in a couple of ways. One way is to nail a 2x4 upright to the gable end ahead of time so that it extends up to the height of the ridge. This allows you to push the end rafters against the upright and to install a 2x4 sway brace extending from the top plate to the ridge at a 45° angle. This is a permanent brace. Nail it in between the layout lines at the ridge.

A second method is to use your eye as a gauge. Sighting down from the end of the ridge, align the outboard face of the end rafters with the outside edges of the top and bottom plates, and then nail up a sway brace. Either way, the ridge can be plumbed without using a level. This means carrying one less tool up with you when you stack the roof.

With the bare bones of the first ridge section completed, raise the remaining ridge sections in the same way, installing the minimum number of rafter pairs and support legs to hold them in place. When you reach the opposite end of the building, eyeball the last rafter pair plumb, scribe the end cut on the ridge (if the ridge is to be cut at the plate line), slide the rafters over a bit and cut the ridge to length with a circular saw. Then reposition the rafters and nail them to the ridge. Install another sway brace to stabilize the entire structure.

Now stack the remaining rafters, installing the frieze blocks as you go. Nail through the sides of the rafters into the blocks, using two 16d nails for up to 2x12 stock and three 16d nails for wider stock. Where a rafter falls next to a joist, drive three 16d nails through the rafter into the joist. This forms a rigid triangle that helps to tie the roof system together.

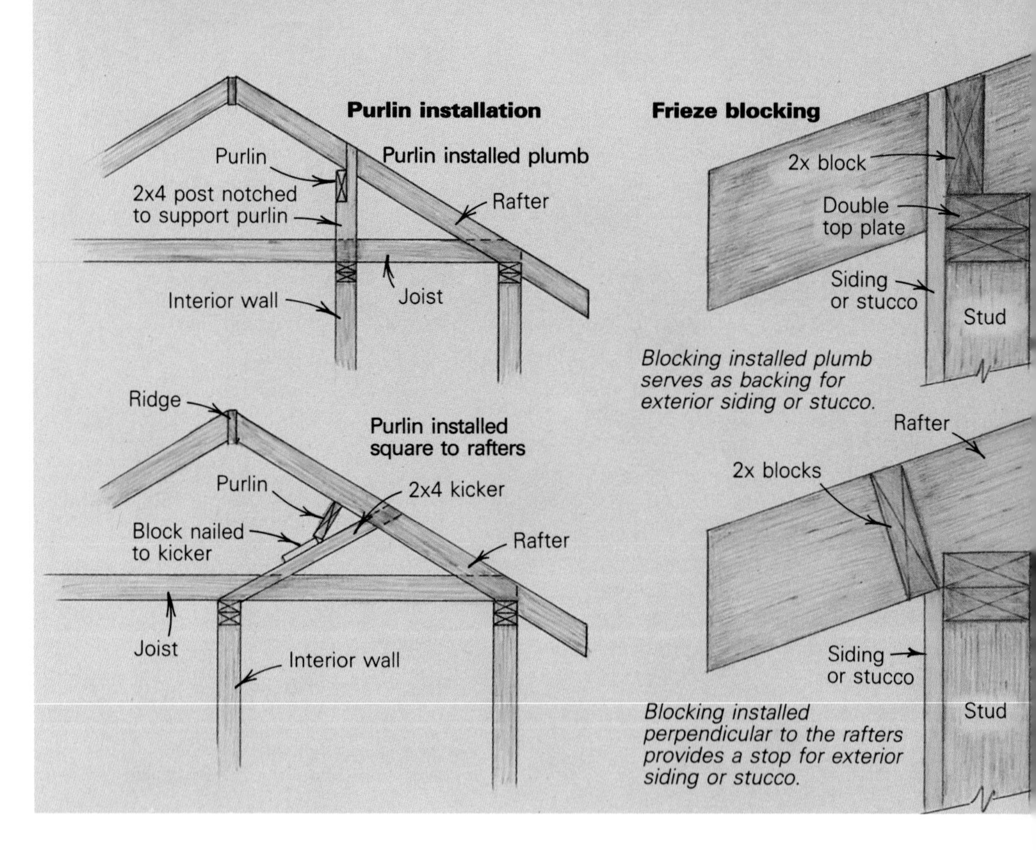

Roof-framing tips

Check your blueprints for the roof pitch, lengths of overhangs, rafter spacing and size of the framing members. But don't rely on the blueprints to determine the span. Instead, measure the span at the top plates. Measure both ends of the building to make sure the walls are parallel; accurate wall framing is crucial to the success of production roof framing.

Once you've determined the length of the rafters, compensate for the thickness of the ridge by subtracting one half the ridge thickness from the length of the rafters. Though theoretically this reduction should be measured perpendicular to the ridge cut, in practice for roofs pitched 6-in-12 and under with 2x or smaller ridges, measuring along the edge of the rafters is close enough. For 2x ridge stock, that means subtracting ¾ in. from the rafter length. An alternative is to subtract the total thickness of the ridge from the span of the building before consulting your rafter book.

Once you've figured the common-rafter length, determine the number of common rafters you need. If the rafters are spaced 16 in. o. c., divide the length of the building in feet by four, multiply that figure by three and then add one more. That will give you the number of rafters on each side of the roof. If there are barge rafters, add four more rafters. If the rafters are spaced 24 in. o. c., simply take the length of the building in feet and add two, again adding four more to the total if barge rafters are called for. *—L. H.*

In some parts of the country, blocking is not installed between the rafters at the plate. But in many areas, building codes require blocks. I think they're important. They stabilize the rafters, provide perimeter nailing for roof sheathing and tie the whole roof system together. They also provide backing or act as a stop for siding or stucco. If necessary, they can easily be drilled and screened for attic vents.

There are two methods for blocking a gable roof (drawings above right). The first is to install the blocking plumb so that it lines up with the outside edge of the top plate, allowing the blocks to serve as backing for the exterior siding or stucco. This requires the blocking to be ripped narrower than the rafters. The other method is to install the blocking perpendicular to the rafters just outside the plate line. The blocking provides a stop for the siding or stucco, eliminating the need to fit either up between the rafters. Also, there's no need to rip the blocking with this method, which saves time. Either way, blocks are installed as the rafters are nailed up. Sometimes blocks need to be cut a bit short to fit right. Rafter thickness can vary from region to region (usually it's related to moisture content), so check your rafter stock carefully.

Collar ties and purlins—In some cases building codes require the use of collar ties to reinforce the roof structure or purlins to reduce the rafter span (drawings above left). Collar ties should be installed horizontally on the upper third of the rafter span. They're usually made of 1x4 or wider stock, placed every 4 ft. and secured with five 8d nails on each end so that they tie the opposing rafters together.

Purlins should be placed near the middle of the rafter span. They can be toenailed to the rafters either plumb or square. If there's an interior wall beneath the center of the rafter span, install the purlin plumb and directly over the wall. This makes it easy to support

Drawings: Michael Mandarano

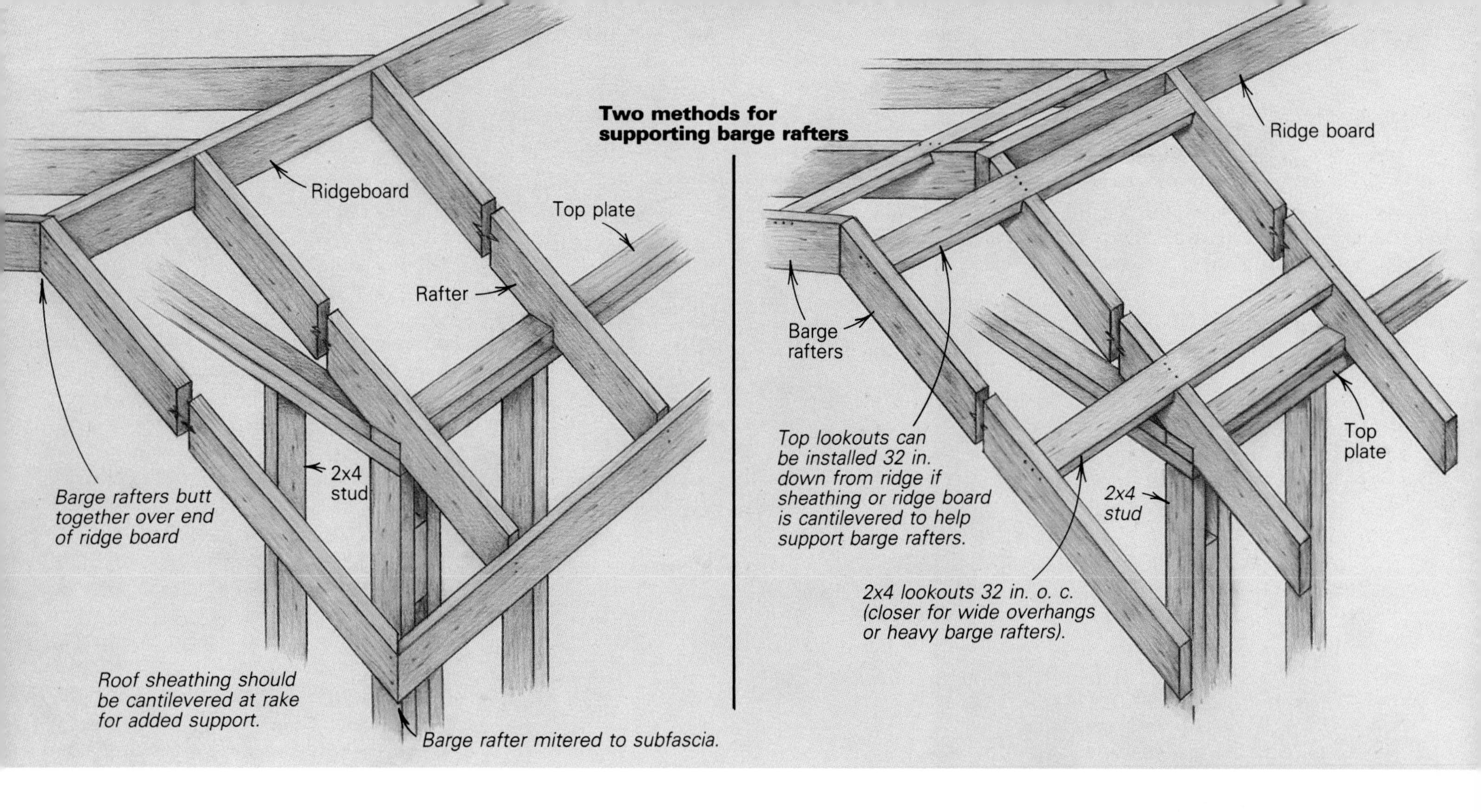

the purlin with several 2x4 posts that bear on the top plate of the interior wall. The 2x4s are notched so that they both support the purlin and are nailed to the sides of the rafters.

If there isn't a wall beneath the center of the rafter span, toenail the purlin square to the rafters and install 2x4 kickers up from the nearest parallel wall at an angle not exceeding 45°. A block nailed to each kicker below the purlin will hold the purlin in place. Kickers are typically installed every 4 ft. Large purlins such as 2x12s require fewer kickers.

In some parts of the country, rafters have to be tied to the top plates or blocking with framing anchors or hurricane ties for added security against earthquakes or high winds. Check your local codes.

Framing the gable ends—Gable ends are filled in with gable studs spaced 16 in. o. c. Place the two center studs (on either side of the ridge) 14 in. apart. This leaves enough room for a gable vent, which allows air to circulate in the attic. Measure the lengths of these two studs, then calculate the *common difference* of the gable studs, or the difference in length between successive studs. Then you can quickly determine the lengths of the remaining studs. A pocket calculator makes it easy.

For a 4-in-12 roof pitch, the equation goes like this: $4 \div 12 \times 16 = 5.33$. Four equals the rise, 12 the run and 16 the on-center spacing. The answer to the problem, 5.33, or 5⅜ in., is the common difference. Another way to calculate this is to divide the unit rise by three and add the answer and the unit rise together. For a 4-in-12 pitch, $4 \div 3 = 1.33 + 4 = 5.33$. For the angle cuts, set your saw to the same angle as that of the plumb cut on the rafters. Cut four gable studs at each length, and you'll have all the gable studs you'll need for both gable ends.

Once the gable studs are cut, nail them plumb using your eye as a gauge. There is no need to lay out the top plates or to align the gable studs with the studs below. Be careful not to put a crown in the end rafters when you're nailing the gable studs in place.

Finishing the overhangs—The next step is to install the *barge rafters* if the plans call for them; these are rafters that hang outside the building and help support the rake. Sometimes barge rafters are supported by the ridge, fascia and roof sheathing. In this case, the ridge board extends beyond the building line so that the opposing barge rafters butt together over its end and are face nailed to it. At the bottoms the barge rafters are mitered to the sub fascia boards, which also extend beyond the building line. The roof sheathing cantilevers out and is nailed to the tops of the barge rafters.

Another way to support barge rafters is with lookouts. A lookout is a 2x4 laid flat that butts against the first inboard rafter, passes through a notch cut in the end rafter and cantilevers out to support the barge rafter (drawing above right). Lookouts are usually installed at the ridge, at the plate line and 32 in. o. c. in between (closer for wide overhangs or heavy barge rafters). If the roof sheathing cantilevers out over the eaves (adding extra support for the barge rafters), then the top lookouts can be placed 32 in. down from the ridge.

The notches in the rake rafters are most easily cut when you're working at the rafter horses. Pick out four straight rafters and lay out the notches while you're laying out for the birds' mouths and ridge cuts. Cut these notches by first making two square crosscuts with a circular saw 1½ in. deep across the top edges of the rafters. Then turn the rafters on their sides and plunge cut the bottom of the notch.

Lookouts are cut to length after they're nailed up. Snap a line and cut them off with a circular saw. That done, the barge rafters are face nailed to the ends of the lookouts with 16d nails.

Pieceworkers don't waste time predetermining the ridge height. Instead, they toenail a pair of rafters to the top plates at either end of the ridgeboard, then raise the ridgeboard between the rafters and nail the rafters to it with 16d nails. A 2x4 sway brace is installed before the intermediate rafters are nailed up.

The final step in framing a gable roof is to snap a line across the rafter tails and cut them to length. Cutting the rafters in place ensures that the fascia will be straight. Use the layout tee or a bevel square to mark the plumb cut. If the rafters are cut square, use a triangular square. Then, while walking the plate or a temporary catwalk nailed to the rafter tails, lean over and cut off the tails with a circular saw. □

Larry Haun lives in Los Angeles and is a member of local 409, where he teaches carpentry in the apprenticeship program. Photos by Robert Wedemeyer except where noted.

Ceiling Joists for a Hip Roof

Three simple, problem-solving framing techniques

by Larry Haun

Heading out the joists

Framing models by Linden Frederick

If you want a simple roof, go for a gable. The angles are basic, and even the ceiling framing is a snap because it generally runs parallel to the rafters. A hip, however, is a roof to reckon with, and not just because of its angles. With rafters sloping up from all four sides of the building, how do you joist the ceiling?

"No problem," you say. But what about the joists that run perpendicular to the rafters on opposing sides of the house? Depending on the roof's slope, it's quite likely that you won't have much room for the joist closest to the plate: the rafters will be in the way. In 41 years of pounding nails, I've seen several solutions to the problem of joisting a hip roof.

Time-honored solutions—Depending on the size of the joists, the ones closest to the plates may need to end at a header to let the hip rafter through (photo left). A similar header is sometimes required for the valley rafter. Backing for the ceiling finish is obtained by nailing 2x blocking flatwise to the plates between the rafters (behind the frieze blocking), to provide 1½ in. of nailing surface for the ceiling drywall.

There are times, however, when the first joist may have to be held away from the plate as much as 32 in., depending on the pitch of the roof and the size of the lumber used for the joists. The traditional method of filling the space between the outside wall and this first joist is to add stub joists at right angles to the main joists, parallel to the jack rafters (bottom photo, facing page). This technique often requires that the first joist be doubled to carry the extra ceiling load. The stub joists are usually clipped on one end to keep them from sticking up above the roof framing; pressure blocks at the other end help to support them.

A better solution—There's another way to handle the problem, one that requires less material and labor. Set your last joist as close as possible to the plate. Then set another one—flatwise—within 16 in. of the outside wall. There will be plenty of room between it and the underside of the hips and jacks. The next step is to install the rafters and to nail frieze and backing blocks between them as described earlier. Then cut strongbacks from scrap 2x stock and run them flatwise from the plate to the first on-edge joist (top photo, facing page). You won't need many of them; spacing them about 4 ft. o. c. should do.

Secure the strongbacks to the backing blocks and to the upright joist with a couple of 16d's at each end. At the joist end the strongbacks must be held up 1½ in. from the bottom. Finally, pull the flat joist up to the strongback, securing it with three or four more 16d's, angling them slightly for better holding power. The strongbacks will stiffen and support the flatwise joist. With this step complete, the ceiling is ready and you're set to move on. □

Larry Haun lives in Los Angeles and is a member of Local 409; he taught in the apprenticeship program. Photos by Susan Kahn.

From *Fine Homebuilding* magazine (August 1991) 69:44-45

Strongback framing

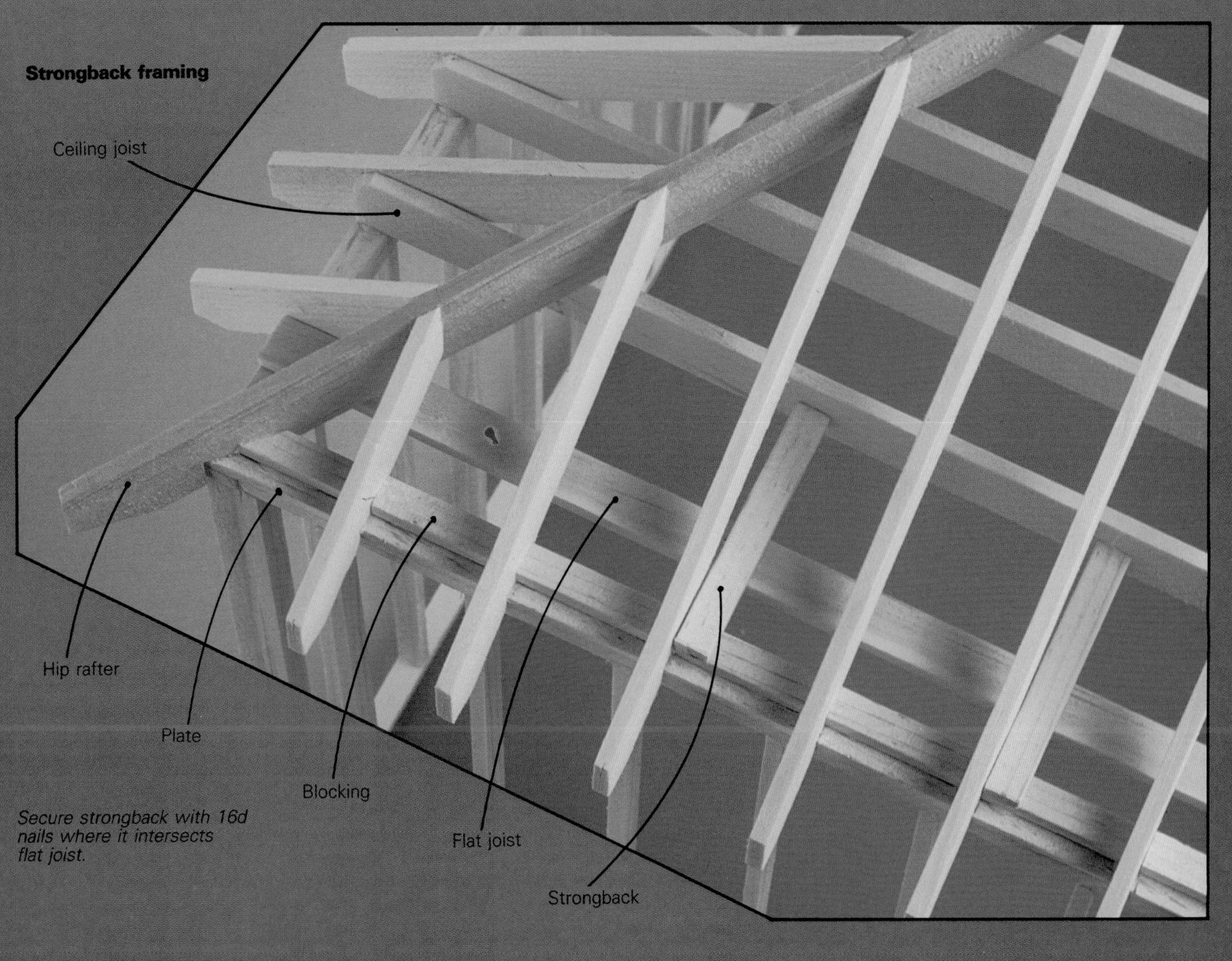

Secure strongback with 16d nails where it intersects flat joist.

Stub-joist framing

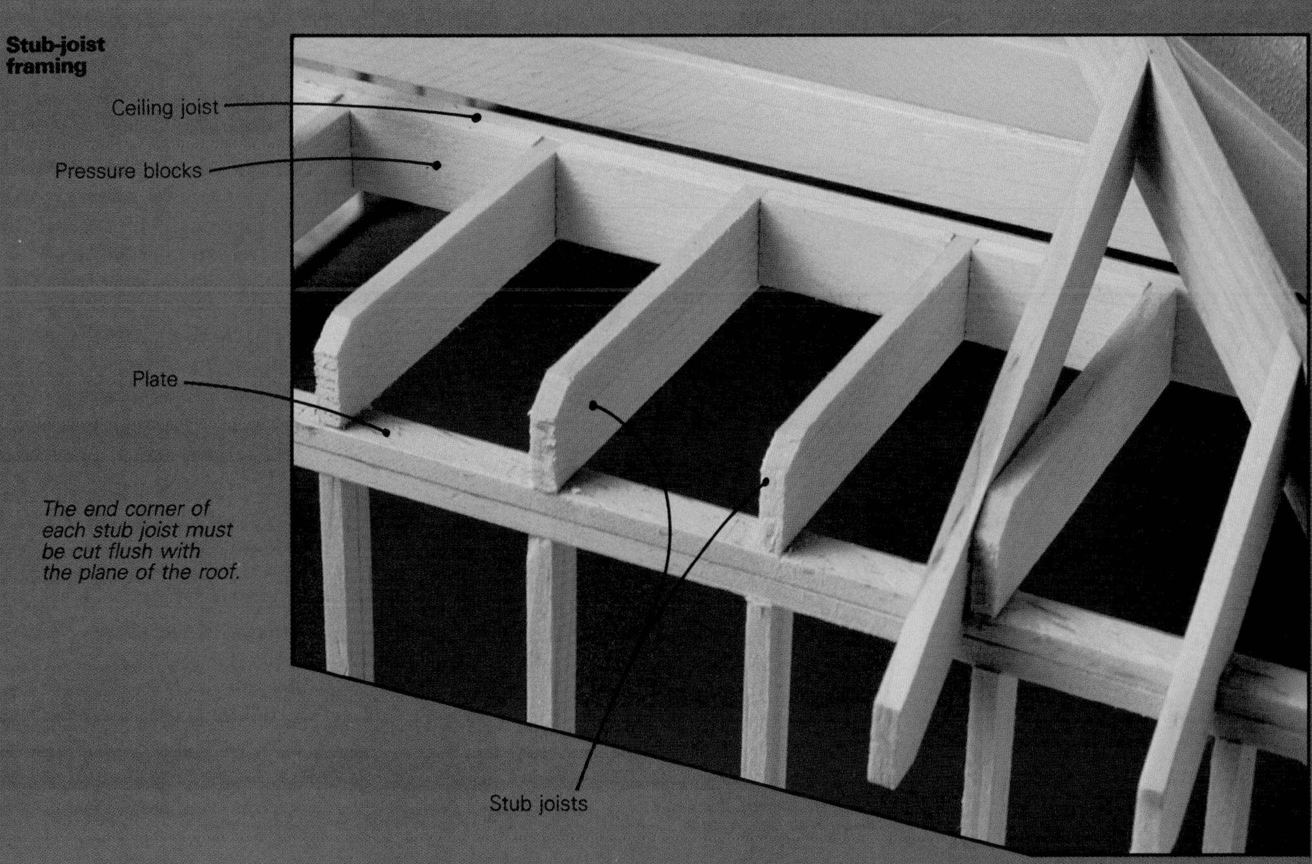

The end corner of each stub joist must be cut flush with the plane of the roof.

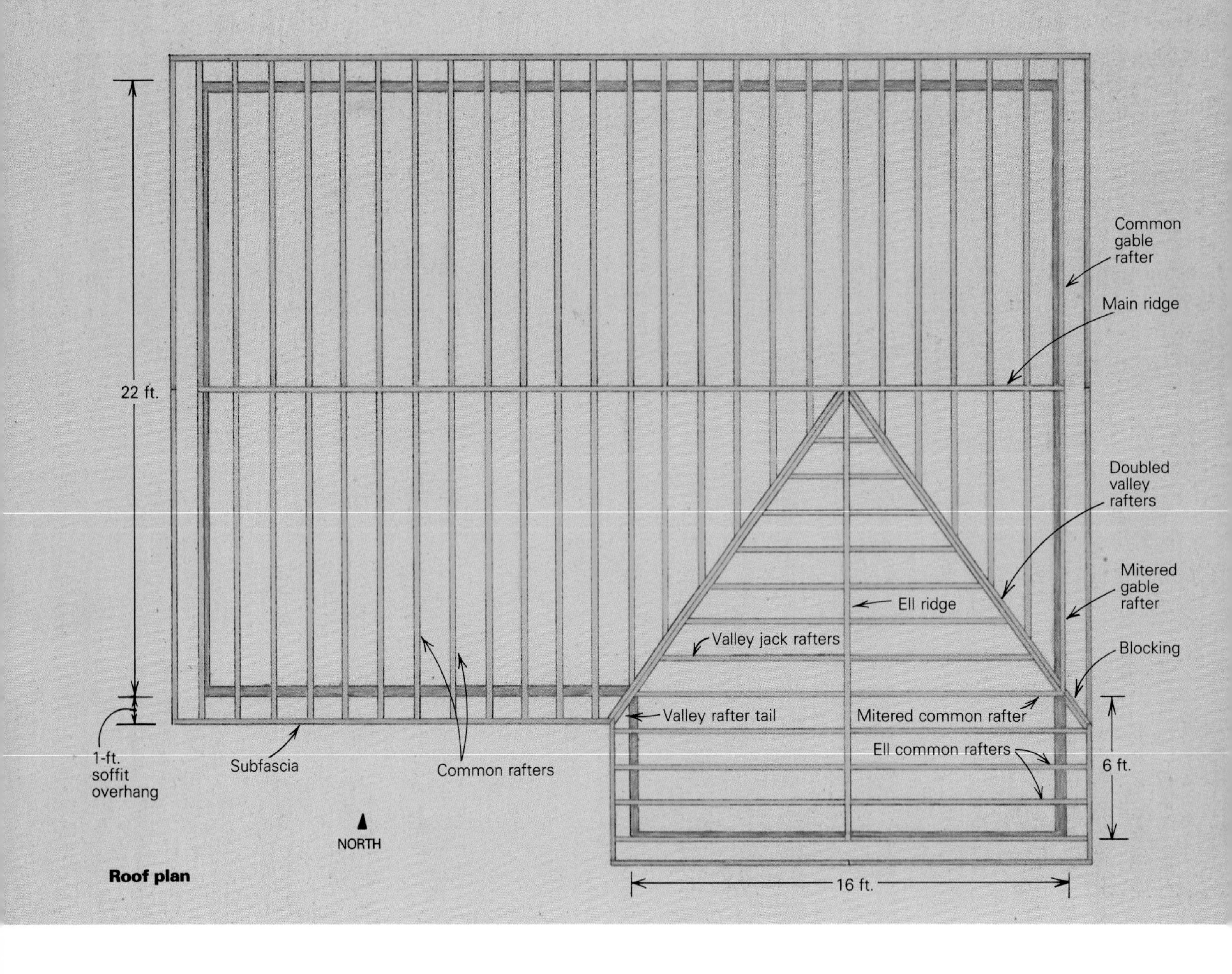

Valley Framing for Unequally Pitched Roofs

An empirical method that works

by George Nash

The intersection of two roofs with unequal pitches involves geometrical relationships not readily visualized or easily understood. Graphic projections like the method detailed by Scott McBride in "Roof Framing Revisited" (*FHB* #28, pp. 31-37) can be intimidating to anyone without substantial framing experience. As for me, I want to frame the roof, not tinker with models, pens, paper and a calculator. In fact, I'm convinced that "fear of ciphering" is so common that few framers have anything more than a vague notion of somehow using "strings and levels" to lay out complex roofs. This sometimes translates into blundering through, trial-by-error or fudge-and-fix techniques, with hopes that the client doesn't show up until the roof sheathing has hidden the mistakes.

That's how it went for me until a summer when everything I built had a weird roof. I needed a framing method that was fast, accurate, relatively simple and, most of all, nonmathematical. I've forgotten all the trigonometry I never learned in high school, so I'm hopelessly doomed to be a string-and-level man. In the article that follows, I'll describe that method for you and apply it to a house I built that's fairly typical of unequally pitched roof framing.

Purists, or those more mathematically adept than I, may find my methods inelegant, or

Drawings, except where noted: Christopher Clapp

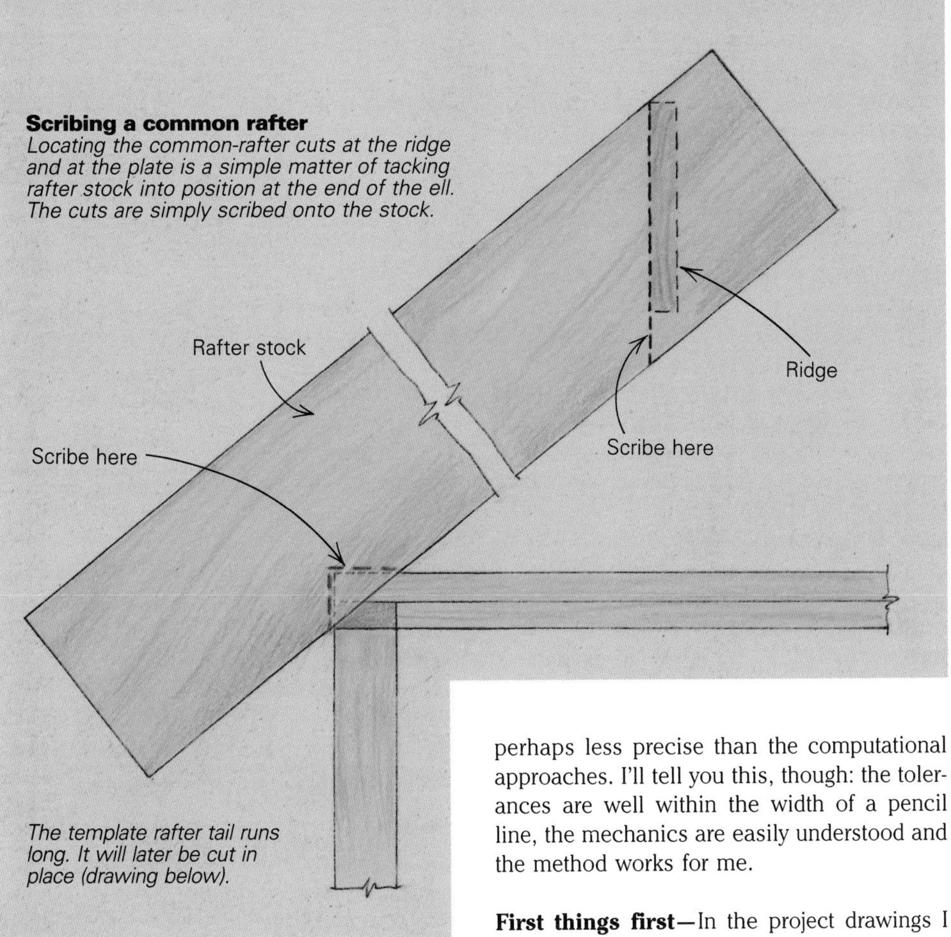

Scribing a common rafter
Locating the common-rafter cuts at the ridge and at the plate is a simple matter of tacking rafter stock into position at the end of the ell. The cuts are simply scribed onto the stock.

The template rafter tail runs long. It will later be cut in place (drawing below).

perhaps less precise than the computational approaches. I'll tell you this, though: the tolerances are well within the width of a pencil line, the mechanics are easily understood and the method works for me.

First things first—In the project drawings I was given, the L-shape of the Stoecklein house appeared to include a conventional valley, but it didn't. According to the drawings, the ridge was at the same height for both roofs, but the rafter span of the main roof was 22 ft. while the ell span was only 16 ft. The main roof was framed at a 7-in-12 pitch. In order for a smaller span to terminate at the same eave height, the pitch of the ell had to be steeper.

The first rule for framing uncommon rafters is to lay out and install all the rafters that ain't (in other words, do the common rafters first). I'll assume you know how to use a framing square to do this; if not check *FHB* #10, pp. 56-61. After cutting and installing all the rafters on the main roof, I was ready to tackle the ell (drawing facing page).

Installing the ridge—As for framing the roof of the ell, the idea is to work from the top down, which means getting the ridge into place and *then* installing the rafters. The first step was to measure and mark the midpoint of the ell top plate and center a plumbed and braced 2x4 post over it. This would support the outboard end of the ell ridge until the ell's common rafters were installed. The post was cut to the same length as the distance between the top plate and the underside of the main ridge. Working from pipe scaffolding, I transferred the centerline of the ell onto the main ridge to locate the intersection of the ell ridge. (I always use rented pipe scaffolding for roof framing. With two sections and enough staging plank, all but the longest roofs can be framed with minimal movement.)

Figuring the length of the ell ridge was easy. It had to run the full length of the ell, plus half of the full width of the main building, minus one-half the thickness of the main roof's ridgeboard. In this case, that meant 6 ft. (the ell) plus 11 ft. (half the main roof) minus ¾ in. (half the ridge). So the ell ridge would be 16 ft. 11¼ in. long. After cutting the ell ridge to length and marking out the rafter spacing on it (better now than when it's up in the air), I nailed it into place at the main ridge and atop the ell centerpost. I double-checked to make sure that the end of the ell ridge ended plumb over the gable wall. A temporary diagonal brace run down to the deck held everything in place.

Ell common rafters—Once the ridges and main-roof common rafters were in place, the layout for the ell common rafters was simple: I pinned rafter stock against the end of the ridgeboard and the corner of the wall plate and scribed for the plumb cut and bird's mouth (top drawing, left). No, it's not elegant, but it works perfectly. The position and depth of the bird's mouth followed from the rule that the seat cut should begin at the inside edge of the top plate. The length of the rafter tail will determine how far away from the wall the fascia will be, so the rafter tails on the ell had to be laid out to allow the ell fascia to flow continuously into the main fascia. Rather than including this step in the initial layout, I simply made the plumb cut at the ridge and the bird's-mouth cuts, leaving ample tail stock to be trimmed later. I cut two rafters and tacked them to the ridge to test the fit.

With the two test rafters in place, it was easy to lay out the cuts on their tails. First I leveled across from the bottom edge of a main-roof common rafter tail to the wall itself, as if laying out a horizontal soffit lookout, and then measured the distance from this mark to the top of the wall plate. Returning to the ell, I measured down the wall this same amount and snapped a level line across the wall. Then it was short work with a level and a pencil to extend this line across the bottom of the extending rafter tails; this would be the level cut (bottom drawing, left). To get the plumb cut, I moved a framing square horizontally across this line until it measured a vertical line equal in length to the plumb cut of the main-roof rafters.

It's important to note that if you want the intersecting ridges to be of the same height and the fascias on both parts of the house to line up, the width of the ell soffit will be less than that of the main soffit. If the ell were wider than the main roof, the reverse would be true. If you'd rather have the soffits be equal in width and at the same elevation all around the house, then one of the ridges must be lowered or raised accordingly. Usually these sorts of details are worked out in the design phase. On this job, the difference in soffit width amounted to slightly less than 3 in., which really isn't noticeable.

Although the method I just described will establish the tail cut for either horizontal or pitched soffits, I'd recommend using a horizontal soffit unless the design is beyond your control or changes are not allowed. Horizon-

Rafter-tail layout

Rafter
Level
Framing square
Offcut
Level line

Determining the length and cutting pattern for a common rafter tail on the ell can be done without calculation. Snap a level line on the wall, corresponding to the common rafter level cut on the main roof. Then use a spirit level to transfer this line to the common rafter tail. Slide a framing square along this line until it measures a given distance along the vertical leg; this will be the plumb cut.

From *Fine Homebuilding* magazine (June 1991) 68:74-78

tal soffit boards and vents are much easier to fit and nail than pitched ones.

About this time I'll usually support the intersecting ridges with a temporary post. Otherwise the ridges could sag as the valleys and their jack rafters are added, and the plumb cuts and lengths of the jack rafters would become increasingly inaccurate. I always check the ridges for straightness, or line them to a string, before laying out the valley rafters.

Finding the valley length—A valley rafter has a lot of cuts and angles to line up, and you'll have a lot of lumber to throw away if one calculation turns out wrong. Fortunately, there's a way to isolate each component and reduce the chances for confusion and error.

Because I always use a subfascia of 2x stock (for the extra support it gives to the soffit), finding the length of the valley rafter isn't tough. First, I nailed the subfascia to all the common rafters around the house. Where the ell intersected the main building, I extended the subfascias to meet at the inside corner (drawings below). Because I had beveled their top edges to match the corresponding roof pitches, it took some fudging with a trim plane to fit the steeper bevel to the shallower one. Some carpenters skip the bevel and simply drop the fascia slightly instead (either method will provide a nailing surface for the edge of the roof sheathing). Then I nailed the intersecting subfascias together. I stretched a string from the outside corner of this intersection to the intersection of the two ridges, right to the top edge; this represented the center line of the valley rafter's top edge (drawings below). Finding the actual rafter length was simple: I just measured along the string.

Figuring the plumb cuts—To find the face angle and the edge angle of the valley-rafter plumb cuts, I used a sliding T-bevel to copy the angle between the string and the ridges. The same angle marked the heel cut of the bird's mouth. It was easy to use a short level and plumb up from the wall plate to the string and then measure the distance to determine not only the heel cut, but the depth to the seat cut of the bird's mouth and the length of the valley from ridge to plate (drawings facing page). The tail cut was simply the same angle repeated where the string crossed the intersecting subfascia. Before making the actual cuts I transferred the angles to a short length of stock and cut a test piece—mistakes on scrap stock are a lot easier to correct.

A doubled valley—A valley rafter on a roof with *regular* pitches calls for a double cheek cut where the valley rafter intersects the ridges and the subfascia. The top edge of the valley rafter then has to be "dropped" just enough to allow roof sheathing to clear it. But the double cheek cut can be complicated, and dropping the valley leaves very little support for fastening the roof sheathing. That goes against the grain of my framing aesthetic—I like plenty of meat to nail into. That's why I double the valley rafter. And if the top edge of each doubled valley rafter is beveled to match the plane of the adjacent roof, the rafter will provide a much better nailing surface for the sheathing (small drawing, facing page).

Of course, two trial pieces with single cheek cuts already made are needed, one for each half of the doubled valley. To find the angle of the top bevels, I lined up each of my trial pieces with the valley rafter center-line string and held it at the intersection of the two ridges. Then I scribed it where the stock projected above the ridge. This is called backing the valley. Because the resulting angle will be scribed across the face of the compound angle (the ridge plumb cut) and not the square edge of the rafter stock itself, it can't very easily be duplicated with a T-bevel. Instead, I used trial and error—when the cut of the table saw matches the scribe line, I've got the right angle. After the top bevels were cut, I installed the paired rafters and spiked them together.

By the way, the same benefits of doubling the valley rafter apply when it must support a finished ceiling. In that case, a 2x4 ripped to the required width and bevel will furr out the underside of the double rafter for solid nailing, and the finished intersection of the different ceiling planes will be more accurate (small drawing, facing page).

While I left a tail on the doubled southwest valley rafters, letting them intersect the fascia, I dispensed with tails on the southeast valley rafters where the ell shared a common wall with the main-roof gable. Instead, I cut a 45° miter in the plate end of the main roof's gable rafter and did the same thing with the intersecting common rafter of the ell. This way they'd fit against each other at the outside of the wall plate and automatically give the correct height for the center-line string. In lieu of the valley rafter tail, the soffit was fastened to a lookout, and blocking above carried the edge of the roof deck.

By the way, I left the center-line strings in place until all the valley jack rafters were finished. Even a doubled valley rafter will shift with the push and shove of the jack rafters and the weight of the carpenters as they clamber

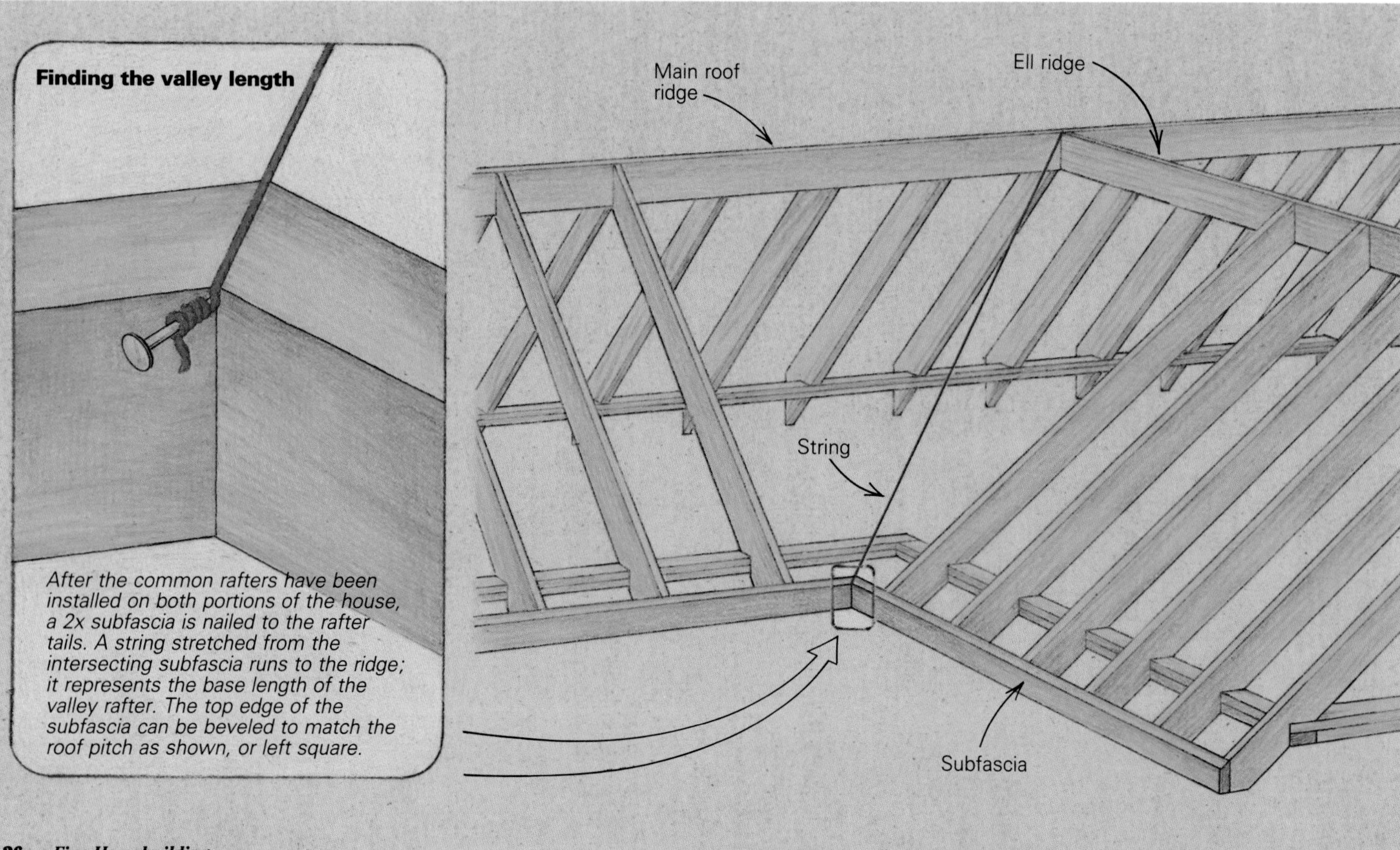

Finding the valley length

After the common rafters have been installed on both portions of the house, a 2x subfascia is nailed to the rafter tails. A string stretched from the intersecting subfascia runs to the ridge; it represents the base length of the valley rafter. The top edge of the subfascia can be beveled to match the roof pitch as shown, or left square.

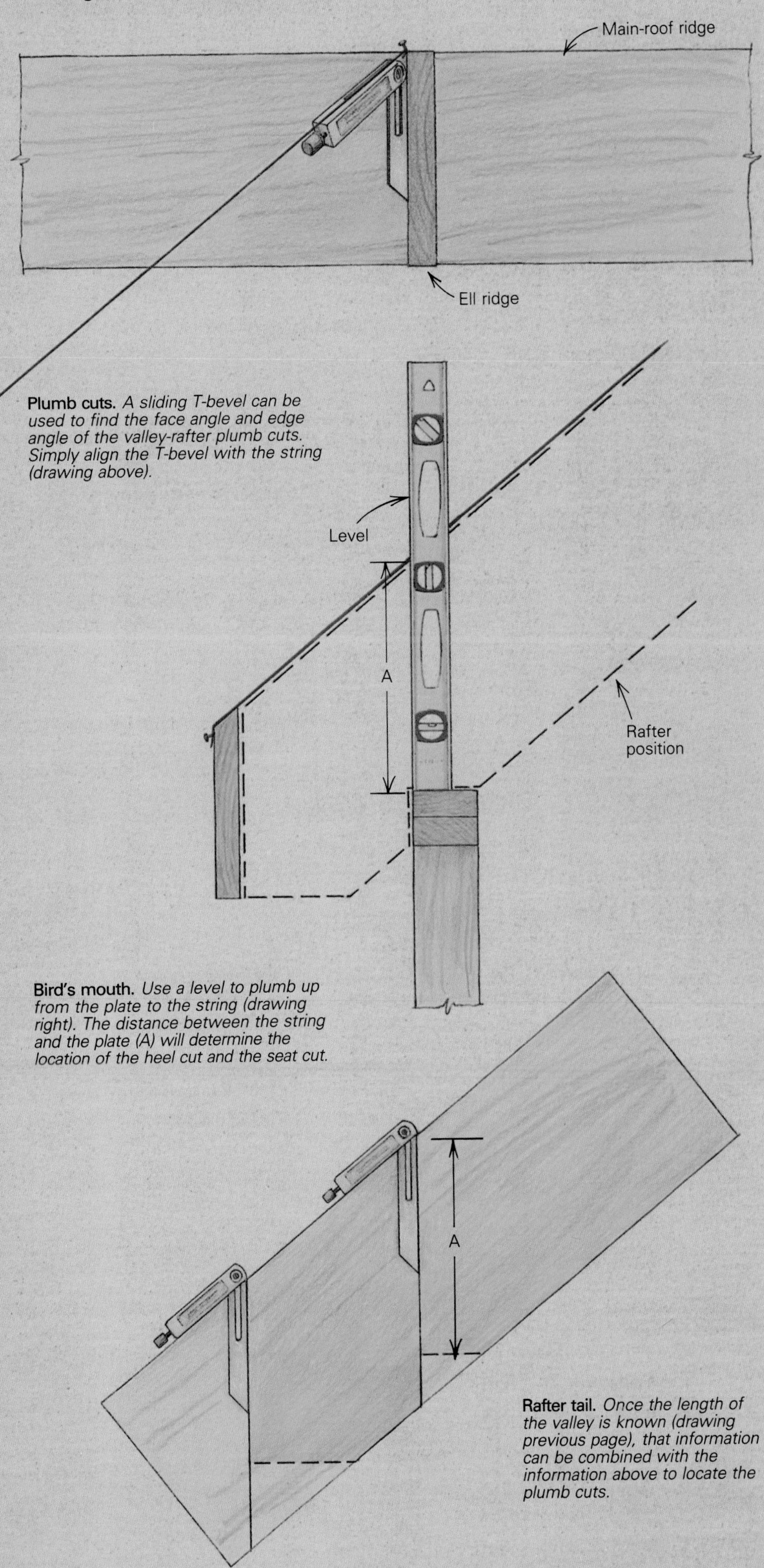

Finding the valley-rafter cuts

Plumb cuts. *A sliding T-bevel can be used to find the face angle and edge angle of the valley-rafter plumb cuts. Simply align the T-bevel with the string (drawing above).*

Bird's mouth. *Use a level to plumb up from the plate to the string (drawing right). The distance between the string and the plate (A) will determine the location of the heel cut and the seat cut.*

Rafter tail. *Once the length of the valley is known (drawing previous page), that information can be combined with the information above to locate the plumb cuts.*

about, especially if the span is long. The string is a convenient guide for constantly checking alignment. Temporary braces may be needed to hold the valley rafter to the line until all the framing is complete.

The valley jack rafters—I have found that the tables of common differences for jack and hip rafters on lines 3 and 4 of the framing square don't always lead to perfect cuts. There are just too many 16ths and smidgens in a real framing job for it to correspond exactly with a theoretical frame. And because I was dealing with an odd pitch on this project, I wanted to derive the common difference (the uniform difference in length between each successive jack) by measuring the actual distance between the first two jack rafters, not by consulting a table. It was string and level time, phase II.

The layout lines for the jack rafters were already marked on both ridges. All I had to do was make a corresponding tick mark on the valley rafter at the right place to find the length and face angle of the jack. I knew that the center of the jack would have to be 16 in. away from the center of the nearest common rafter and be parallel to it, so I was able to use my square to pinpoint its intersection with the valley. (You'll have to eyeball the common rafter for straightness and take out any bows by bracing with a temporary board before you try this.) With the long side of the square resting on top of the common rafter and the short side resting on the valley rafter, I simply slid the square up and down until I located a point on the outside of the valley exactly 16 in. away from the outside of the common rafter (top drawing, next page). This represented the intersection of the valley jack with the valley.

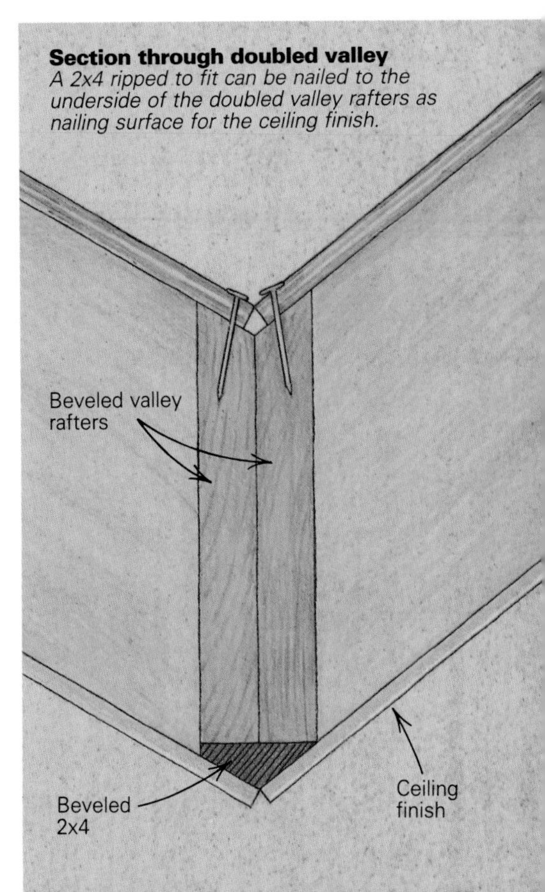

Section through doubled valley
A 2x4 ripped to fit can be nailed to the underside of the doubled valley rafters as nailing surface for the ceiling finish.

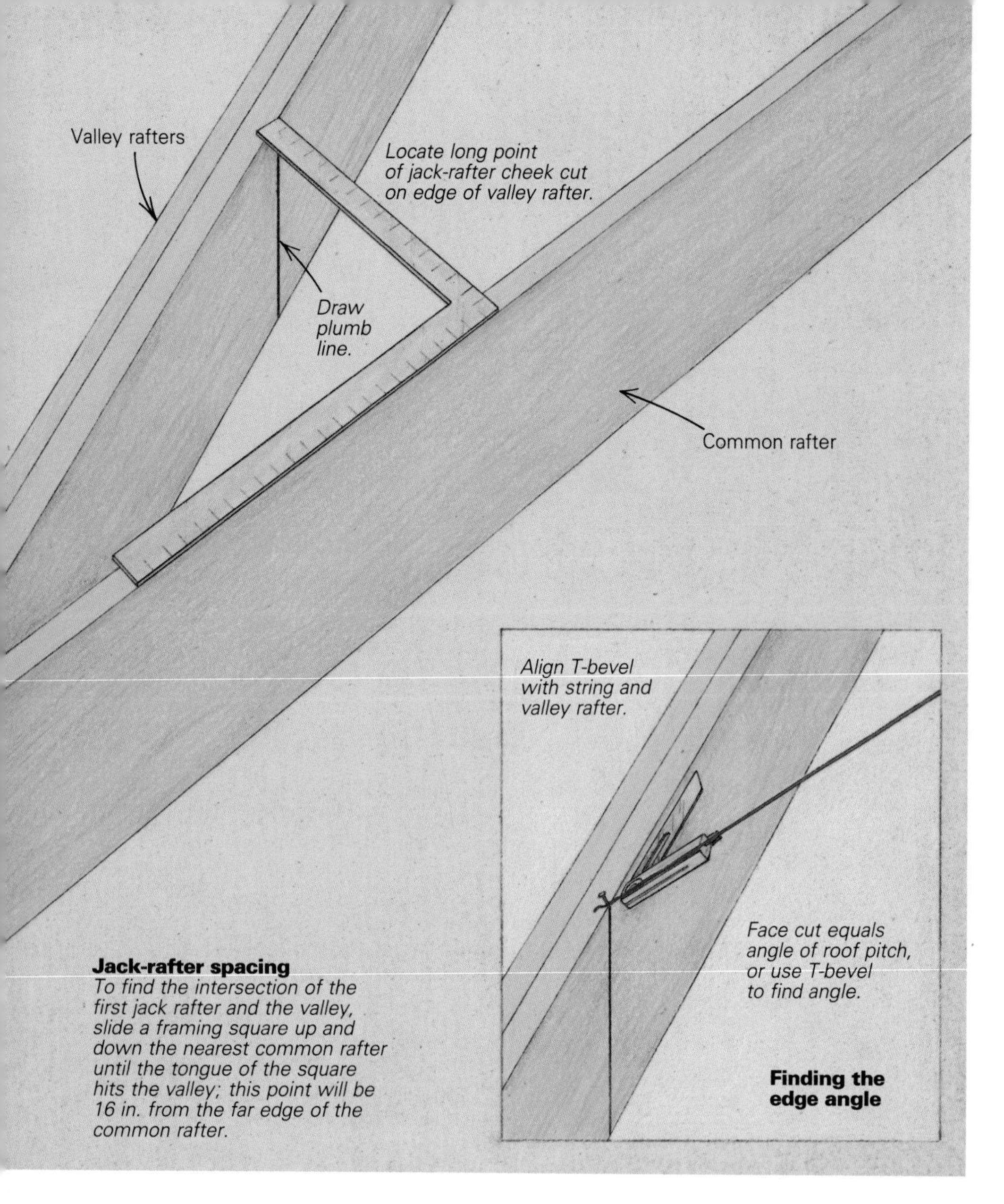

Jack-rafter spacing
To find the intersection of the first jack rafter and the valley, slide a framing square up and down the nearest common rafter until the tongue of the square hits the valley; this point will be 16 in. from the far edge of the common rafter.

Then I drew a plumb line down the side of the valley rafter using a short level.

I fastened a string from the top of this mark to the ridge, parallel to the common rafter, and stretched it tightly. Then I aligned the body of the T-bevel with the string and set the blade against the valley rafter; this gave me the angle of the cheek cut. It's important that the blade and the center of the T-bevel handle be in the same plane when lined up to the string; a small twist will give you an incorrect angle. Here again, you'll want to make a trial piece from scrap stock to test the fit before cutting the actual rafter. Because the valley had been doubled up earlier, instead of dropped, the point marked was exact.

To determine the jack-rafter length I simply measured the string from the ridge to the tick mark on the valley rafter. I duplicated the angle of the plumb cut at the ridge on each successive piece using a framing protractor. Making the compound cuts first before laying out the plumb cut at the ridge end ensures a good fit. The string was no longer needed once I found the angles and verified the fit.

After I installed the first jack rafter, it was easy enough to determine the common difference simply by measuring it. All I had to do now was repeat the series of cuts on each remaining jack rafter, reducing the length of each one by the common difference.

To keep the valley rafter in line, the jack rafters are usually nailed home in opposing pairs. But with unequal roof pitches, the cheek cuts are not mirror images and the spacing intervals will not line up across from each other. To avoid throwing subsequent measurements off, I find it easier to work up one side of the valley at a time, marking the position of the next jack rafter by setting the framing square along the edges of the last. I used bracing to keep the jacks from crowding the valley rafter off the center line.

The beauty of this empirical method is that no advance preparation is required before framing can begin. A good rule of thumb is: if you count the jack rafters for one side of the valley as if they were common rafters and add an extra, there won't be much waste. The shorter jacks are usually cut from the leftovers of the longer ones. □

George Nash lives in Burlington, Vt.

Making cheek cuts

The worst thing about jack rafters is making the cheek cut, which is almost always greater than 45°. Sawing through a 2x12 at a compound angle with a handsaw is tedious and tiresome; using a chainsaw is dangerous and usually not very accurate, and I don't have a compound miter saw. Instead, I used applied geometry, some power and a dash of old-fashioned elbow grease.

For example, suppose the edge angle figures out to be 72° and the face angle 35°. I first cut the face angle across the face of the rafter (with the saw set at 90°) and then tack the rafter to the sawhorse. Complementary angles must add up to 90°, so the complement of 72° is 18°. If I set the saw at that angle and then hold its base against the edge cut itself (perpendicular to the side of the rafter) I can make a 72° cut. Although a 7¼-in. blade will not cut all the way through the angle, what's left is fairly easy to finish with a handsaw. An 8¼-in. or 12-in. circular saw would be handier. I know of no easier way to make these cuts than on a radial-arm saw, which takes more time to set up.

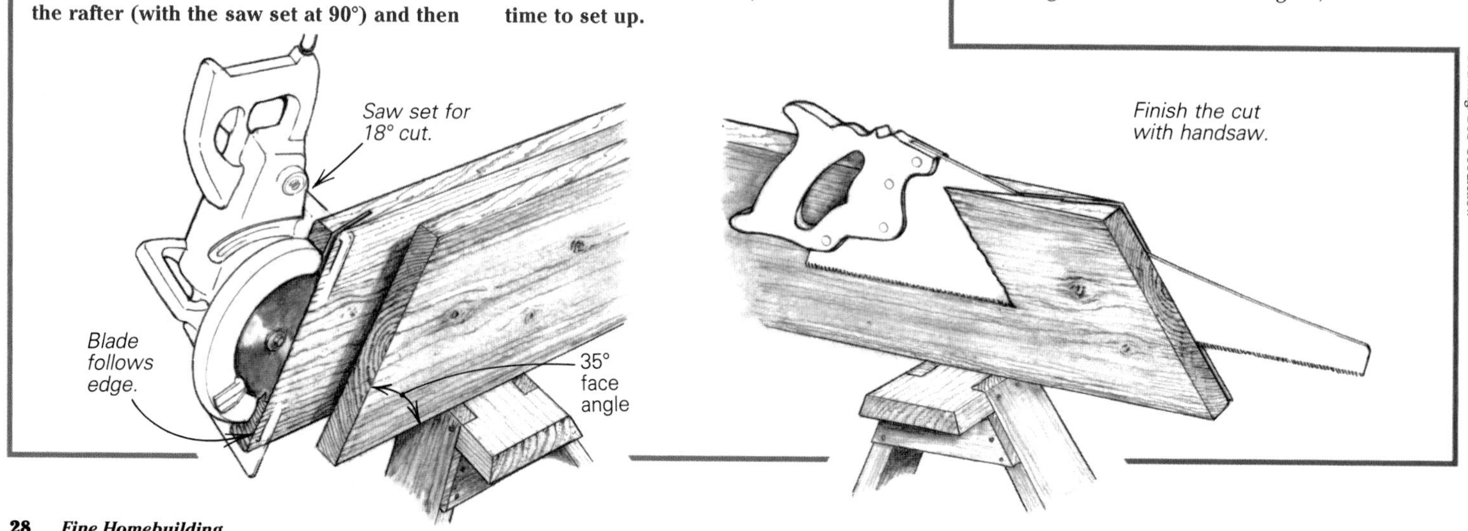

Drawing: Bob Goodfellow

With its cross-gabled roof and decorative vergeboards, this cottage complements the Gothic home to which it's appended. Photo by Kevin Ireton.

Framing a Cross-Gable Roof

One good valley rafter supports another

by Scott McBride

In 1851 a German immigrant named Henry Kattenhorn owned a thriving sugar refinery in the riverfront village of Hastings-on-Hudson, New York. Deciding that his four superintendents and their families should share in his prosperity, Kattenhorn built cottages for them on a bluff overlooking the river. Bedecked with finials, decorative chimneys and gaily sawn vergeboards, these small, cozy houses were prime examples of Gothic Revival architecture.

Just a year before the Kattenhorn cottages were built, Andrew Jackson Downing, the leading exponent of the Gothic Revival style, had published *The Architecture of Country Houses*. In his book, Downing had inveighed "an excess of fanciful and flowing ornaments of a cardboard character," but the country carpenters who adapted the style from readily available pattern books were hard to restrain—lumber was cheap, the steam-driven jigsaw had been invented, and the sky was the limit.

Besides gingerbread, another hallmark of the Gothic Revival style was the cross-gable roof. Downing also tried to temper the proliferation of gables, lamenting that "some uneducated builders...have so overdone the matter, that, turn to which side of their houses we will, nothing but gables salutes our eyes." But the "cocked-hat cottage," as Downing called small dwellings with multiple gables, was precisely the form chosen for a recent addition to one of the Kattenhorn cottages (photo above).

When Judy Seixas approached architect Stephen Tilly about adding a semi-detached bedroom suite to the back of her house, she was adamant that the design be strictly in keeping with the Gothic Revival style. Tilly and chief designer Laurel Rech came up with a simple cross-gable rectangle for the addition. An existing flat-roofed screen porch would be enclosed to house a bathroom, the utility room and an entrance foyer. The converted porch would also link the bedroom suite to the existing house. I was hired to build the addition, the trickiest part of which turned out to be framing the cross-gable roof.

Blind valleys—My crew framed partitions in the former screen porch while the foundation for the new addition was being built. As the blockwork was finished and floor framing began, I retired to a shady spot on the driveway to lay out and cut the principal roof members.

In a pure cross-gable roof, two ridges—both at the same elevation—intersect at 90°. All four valleys formed by the intersection converge at a central peak. Our addition would be a modified version, insofar as there would be a higher continuous ridge and a slightly lower ridge broken by the intervening higher gable. It could be called a gable with two dormers, except that I think of dormers as being subordinate in size to a main roof. The similar size of all four gables on this roof makes them more or less equal partners in the deal.

I have seen cross-gable roofs in Victorian houses where a lower ridge flies right through the attic space under a higher ridge. But because our addition was to have a finished cathedral ceiling, I broke the lower ridge into two discontinuous sections. I considered supporting these lower ridges by hanging their inboard ends on headers framed between the common rafters of the higher gable, which is how I frame gable dormers. But the lower gables in this case were so broad that we would have needed a 13-ft. header to span the distance, which would have been an impractical arrangement.

We resorted to a supporting valley, or a blind valley. For each of the lower gable roofs, one valley rafter would run from the wall plate to the main ridge (photo, p. 30); this is the blind valley. The other valley would be shorter and would intersect the blind valley. This intersection marks the terminus of the lower ridge.

Because the addition's plan was symmetrical, it didn't matter which valley of a pair would run through to the ridge. But I did decide to make the blind valleys from opposing sides of the roof come together at the same point on the ridge. This way, any force exerted on the ridge by one

Blind valley. *Two factors complicated the framing of this roof: One pair of gables is lower than the other; and the room below gets a cathedral ceiling. The solution was to run one valley rafter on each side of the roof through to the main ridge. This is called a blind valley, and it carries the shorter valley and the lower ridge. The framing plan below shows a bird's-eye view of these parts.*

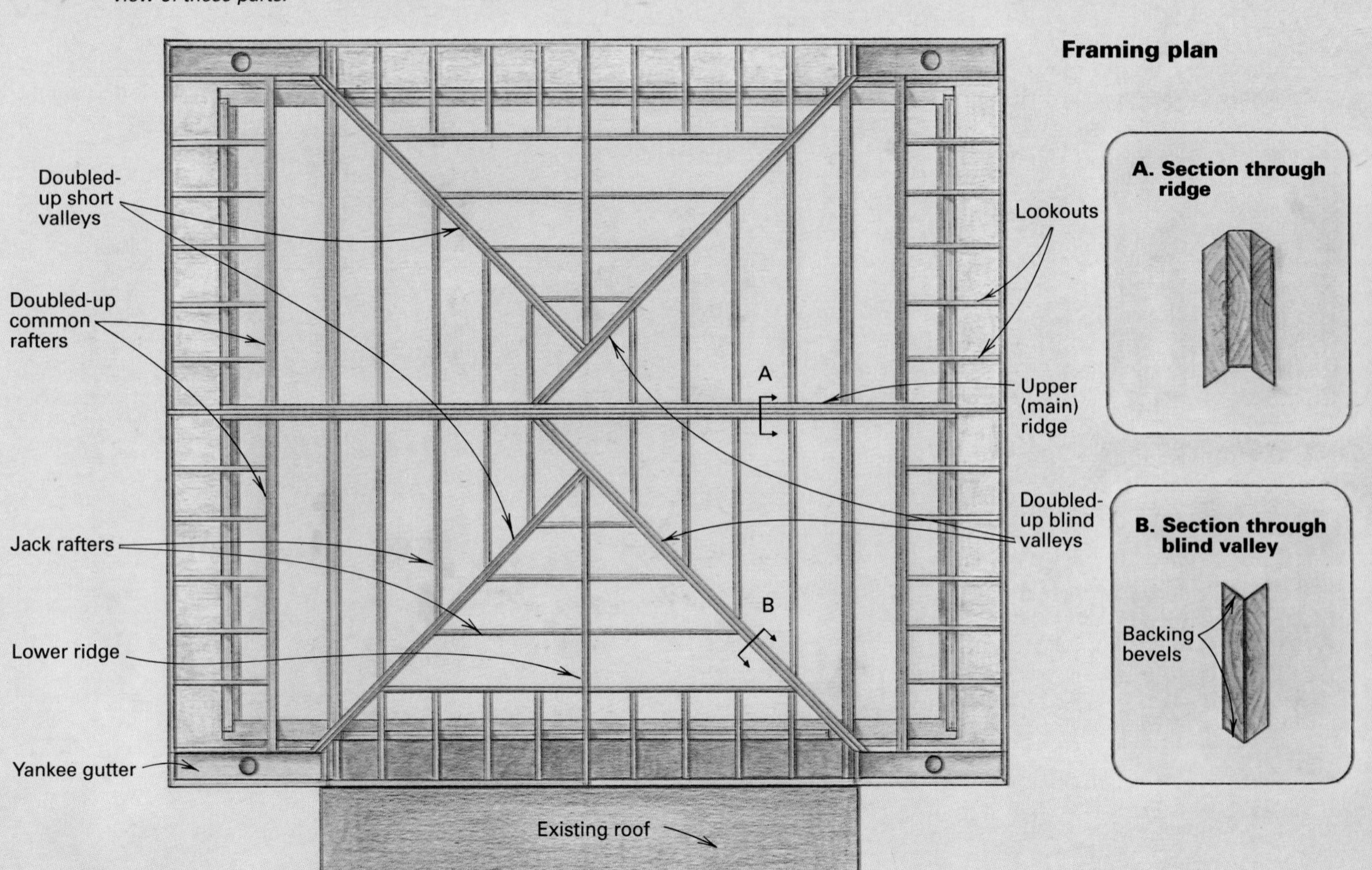

Drawings: Bob Goodfellow

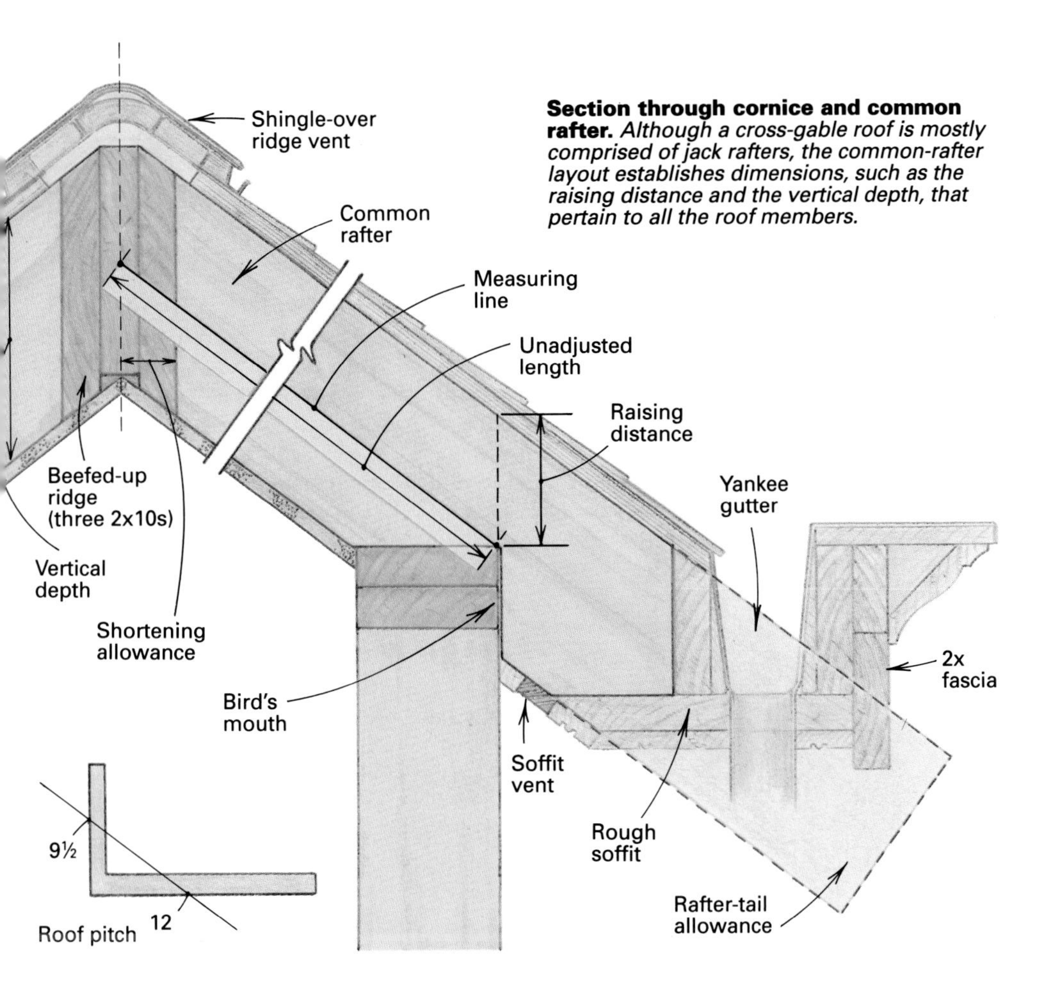

Section through cornice and common rafter. *Although a cross-gable roof is mostly comprised of jack rafters, the common-rafter layout establishes dimensions, such as the raising distance and the vertical depth, that pertain to all the roof members.*

blind valley would be canceled out by the force of the opposing blind valley.

Initially, we didn't want collar ties piercing the cathedral ceiling (although we added a pair, which I'll tell you about later). To compensate for the structural loss of the collar ties and to support the weight of the valleys, we beefed up the main ridge. Three 2x10s spiked together became a structural ridge beam. Because of the girth of the ridge, I beveled the top and bottom edges of the two outside 2x10s so that they wouldn't interfere with roof and ceiling planes.

The ridge beam was the first roof member to go up. We supported it on the end walls of the higher gable and put a temporary post under the spot where the blind valleys would meet.

Common-rafter layout—Because the ridges were at right angles to each other, and the roofs were the same pitch (9½-in-12), I was dealing with regular roof framing, meaning that the compound edge bevels on all my valley and jack rafters would be cut with my circular saw set at 45°. Knowing this, I decided to forego the graphic-development method I use to lay out complex, irregular roofs and resorted to more direct numerical methods.

On a clean sheet of plywood, I laid out a full-scale section of the cornice (drawing above). Next I drew in the top edge of the 2x8 common rafter and the measuring line, which is parallel to both edges and originates at the *outside* corner of the plate. The distance along an imaginary plumb line reaching from the measuring line to the top edge of the rafter is the *raising distance*—a key measurement that would remain constant for all the commons, the valleys and the ridges in the roof frame.

To figure the rafter-tail length, I referred to the blueprints. The architect had furnished me with a wall section showing a copper-lined Yankee gutter, which was to be recessed into the roof at the four short sections of eaves located at each corner of the addition. I couldn't envision how this cornice would return into the vergeboard of the higher gable at one end or how the valley flowing into the gutter would be resolved at the other end. I decided to play it safe by letting the valleys and the commons run long by a generous amount, figuring I'd trim them when I could see things in three dimensions.

After drawing the cornice section, I had the information I needed and could then transfer that information to the rafter stock. Laying a piece of rafter stock in front of me, I scribed my measuring line down its length, offset from the rafter's top edge by the raising distance noted earlier. From the end of the 2x8, I measured up 2 ft. for the rafter-tail allowance and drew a plumb line. Then I drew a level seat cut through the intersection of the plumb cut and the measuring line. I now had my bird's mouth. From the corner of the bird's mouth, I measured the unadjusted length of the rafter along the measuring line.

If the rise of the roof had been a whole number, such as 5-in-12, I could have found the rafter's length in a rafter table such as the one found stamped on the blade of my framing square. But because the pitch was 9½-in-12, I fell back on my trusty Construction Master calculator (Calculated Industries, 22720 Savi Ranch Pkwy., Yorba Linda, Calif. 92687; 800-854-8075). The Construction Master "speaks" in rise-per-foot rather than in sine/cosine, so you don't have to know trigonometry to use it. I came up with an unadjusted rafter length of 10 ft. 2$\frac{7}{16}$ in.

So at a point on the measuring line 10 ft. 2$\frac{7}{16}$ in. from the corner of the bird's mouth, I drew a plumb line representing the unadjusted length of the common. To compensate for the 4½-in. thickness of the ridge, I drew another plumb line back from the unadjusted length by a 2¼-in. shortening allowance—half the ridge thickness.

As a result of the cross-gable configuration, only the first rafters in from each corner of the main roof were commons; the rest were jack rafters. These commons would anchor the lookouts for the gable overhang, though, so I doubled them.

Valley layout—Before laying out valleys, I prefer to rip the backing bevel on the upper edge of the valley stock (see sidebar, p. 32). I usually bevel two pieces and nail them together later to make a double valley rafter with a V-trough down the middle. The backing bevel helps me orient the compound cheek cuts on both ends of the valley; cheek cuts go either outward or inward in relation to the center face of the valley.

If the ceiling below a valley is a cathedral-type, as was the case here, the lower edges of the valley stock should be beveled as well, with upper and lower edges parallel to each other. This keeps the underside of the valley rafter flush with the underside of the jack rafters and makes it easier to install the drywall. The vertical depth of the valley on both faces should be the same as the vertical depth of the commons and the jacks. (Vertical depth is the width of the rafter as measured along a plumb line.)

After ripping backing bevels on all the valley stock, I started laying out the first blind valley. I designated a top edge and a center face with a lumber crayon. At some arbitrary point on the center face, I drew a plumb line using the numbers 9½ and 17 on the square. I used 17 instead of 12 for the unit run because regular hips and valleys always run 17 in. diagonally for every 12 in. that the corresponding common runs perpendicular to the plate. The rise (in this case 9½) remains the same.

Measuring down from the top edge along the plumb line, I laid off the same raising distance I had found for the common rafters. Through the resulting point, I scribed a measuring line parallel to the rafter's edge. Starting at one end of the rafter, I laid off along the measuring line an allowance for the rafter tail. I had to leave more tail length for the valley rafter than for the common rafter because the valley tail, like the valley rafter, would have a greater run.

Once the tail allowance was established, I drew the bird's mouth with its corner on the measuring line, using 9½ and 17 on the square for plumb and seat cuts. From the corner of the bird's mouth, I laid off the unadjusted length of the blind valley. I got this number by using the HIP/VALLEY key on my Construction Master. With this key, I converted the length of the common rafter to the length of the valley rafter.

From the unadjusted length, I stepped back one half the thickness of the ridge measured at 45°. In this case, the diagonal thickness of the 4½-in. thick ridge was 6⅜ in., so I pulled the actual plumb cut back half that, or 3$\frac{3}{16}$ in., from the unadjusted length. This adjustment, like all

Two methods of finding backing bevels

Beveling the top and bottom edges of a valley (or hip) rafter keeps them coplanar with the roof and the ceiling, which simplifies the installation of roof sheathing and drywall. I use two methods to find the backing bevels of hips and valleys: the scrap-block method and the graphic-development method. — *S. M.*

***Scrap-block method*—To use the scrap-block method, begin by bisecting the angle formed by the adjoining walls where the rafter will sit. In the case of regular roof framing, that means bisecting a 90° corner at 45°. You can do this on the actual plates, but I usually just draw on a sheet of plywood or a piece of paper a 90° corner with a 45° line running through it.**

Next, I cut a scrap block with the level seat cut of the valley at the lower end and a square cut at the upper end. The block doesn't have to be the same width as the actual valley, but it must be the same thickness. Set the block on the drawing, with its point on the vertex (photo below left). If the valley is a single 2x, the block should straddle the bisecting line, with its two faces offset ¾ in. to either side. If the valley is to be doubled, set one face of the block on the bisecting line.

From the point where the outside face of the block crosses the plate line, scribe a line on the face of the block parallel to the block's edge. This line indicates the downhill side of the backing bevel. The uphill side of the bevel will be either a center line drawn down the top edge of the block (in the case of a single 2x valley rafter) or the upper corner on the opposite face of the block (in the case of a double valley rafter). The angle is the same in either case.

On the end grain of the square cut, connect the downhill side of the backing to the uphill side (photo below right). This is the ripping angle for your circular saw.

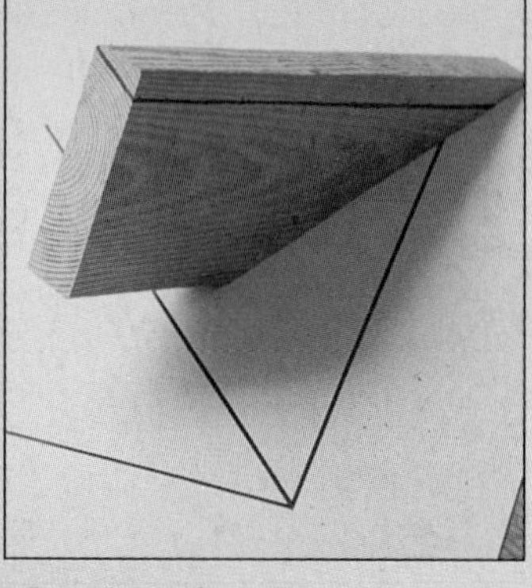

***The scrap-block method.* Beveling the edges of a valley rafter makes it easier to nail on plywood and drywall. Here, a scrap block cut with the seat cut of the roof pitch is used to find the correct angle.**

***Graphic-development method*—The graphic-development method is the same process as the scrap-block method, but it's performed in two dimensions. Suppose the pitch of the valley is 9½-in-17. Starting with the same plan—a 90° angle bisected by a 45° line—apply a framing square with 17 at the vertex and 9½ on the bisecting line (drawing below). Scribe along the 17 side. This is the slope line and is essentially a view of the scrap block pushed over on its side. Next, draw a perpendicular line at any point along the slope line, until it hits the bisecting line. This is a side view of the square cut you made on the scrap block.**

Where the perpendicular line hits the bisecting line—point X—extend lines perpendicular to the bisecting line in both directions until they hit the plate lines at points A and B. The distance from A to B is analogous to the thickness of the valley. With a compass at point X, swing the original perpendicular line down in an arc to hit the bisecting line and connect the resulting point C with A and B. Now imagine you're looking at the end grain of the square cut you made on the scrap block; lines BC and AC represent the lines you drew connecting the downhill side of the backing bevel to the uphill side. Angle ACX is the circular-saw tilt angle.

For an irregular plan, when the walls intersect at some angle other than 90°, the bisecting line will not be 45°. Otherwise, the procedure for finding the backing bevels is the same.

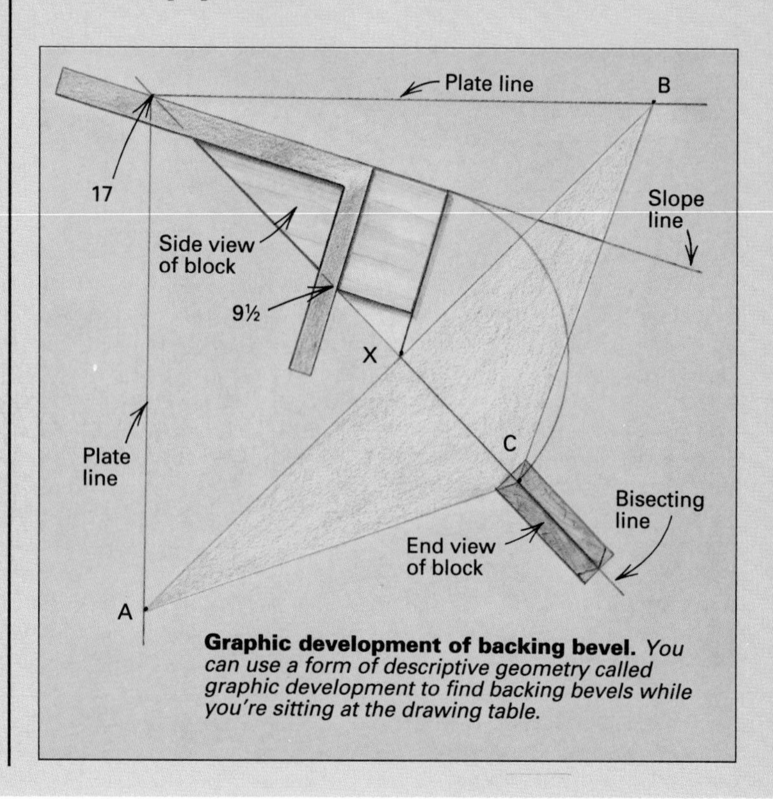

Graphic development of backing bevel. *You can use a form of descriptive geometry called graphic development to find backing bevels while you're sitting at the drawing table.*

shortening adjustments, was made in a horizontal direction, not along the measuring line.

I now had the valley rafter's true length, but I still needed to ascertain the direction of the bevels for the rafter's two compound plumb cuts—the first one located at the back of the bird's mouth where it would fit up against the edge of the plate, and the other at the top of the valley where it would bear against the ridge. After checking the plan, I looked down on the edge of the valley rafter and visualized its position in the completed frame. I then made crayon marks to indicate whether the bevels would go inward or outward from the center face. I cut one half of the double valley and used it as a template for the other half, being careful to orient the bevels in their correct relationship to the center face; the two halves were opposite in this regard.

I cut another pair of rafter halves for the blind valley on the other side of the roof. This pair was the mirror image of the first, with the bevels going in the opposite direction. When both pairs of blind valleys were cut and nailed together, we hauled them up to the roof for the acid test. I got a lot of grief from the crew for all my ciphering, so I was relieved when both valley rafters dropped perfectly into place.

Short valleys, low ridges and jacks—The short valleys were laid out in much the same way as the blind valleys into which they would butt, though with a few differences. Their length was extrapolated via calculator (the HIP/VALLEY key again) from the length of the lower-gable common rafter instead of from the upper-gable common rafter. This relationship is evident in the framing plan on p. 30.

You can also see from the plan that the short valley butts squarely into the blind valley, which seems peculiar if you're used to the oblique orientation of most valleys. Consequently, the plumb cut at the top of the rafter was made with the saw set square, as for a common, and the shortening allowance was half the thickness of the blind valley rafter measured at 90° (not the 45° thickness).

The inboard ends of the short lower ridges, where they nuzzle into the intersection of the

A slight adjustment. **Where the blind valley extends above the lower ridge, the backing bevel had to be reversed on one side (the left side in the photo above) so that it wouldn't break the plane of the roof. The author scored this section of the rafter with a saw and used an ax to hew it flush with the roof.**

A little insurance. **To strengthen the connection between roof and walls, steel brackets were added between the top plates and the longest jack rafters, which were doubled up and were the only rafters with collar ties.**

Cathedral ceiling. **The careful framing of the roof makes possible the crisp lines of a cathedral ceiling. Exposed collar ties keep the walls from spreading. Photo by Kevin Ireton.**

blind valley and the short valley, got a double 45° bevel cut made square across their faces. Once the short valley rafters and lower ridges were nailed in place, I had to make an adjustment to the blind valleys. Where the blind valleys extend above the lower ridge, the backing bevel had to be reversed on the side closest to the short valley. I scored it with a saw and used an ax to hew it flush with the main roof (top left photo, above).

A cross-gable roof is mostly jack rafters, or "fill" as they're referred to on the West Coast. Although methods exist for cutting sets of jacks in diminishing progression, I find that the accumulation of error produced by this system makes it more trouble than it's worth. I just lay off the positions of the jacks on the ridge and on the valley, then measure in between. With a lumber crayon, I scribble the measurements on the ridge beam large enough to read them from the ground.

If the addition's plan had been a square, the outward thrust of the valley at each corner would have been resisted by a pair of walls perpendicular to each other. If the walls were adequately tied together, neither collar ties nor structural ridges would have been necessary. But the plan was rectangular, and we worried about the valleys pushing against the long walls a couple of feet in from the corners. A stronger ridge was one alternative, but strengthening the ridge would have been difficult without making it deeper and bringing it below the ceiling planes. So to tie the opposing long walls together, we bolted clear fir collar ties between the longest pair of jack rafters at each end (photo above right). We used steel angle brackets, cut from heavy angle stock, to strengthen the connection between the jacks and the walls (bottom left photo, above).

Vents and vergeboards—Venting a cross-gable roof that has a cathedral ceiling is problematic because there's little or no eaves soffit to provide cool-air intake. We had only one bay at each end vented at the eaves, but by taking a notch out of the top edge of each jack rafter toward its lower end, we managed to get at least a little draft in the bay's bordering valleys. I also could have recessed the top edge of the valley in relation to the jacks, as I sometimes do with hips, but this would have reduced its strength.

We vented the ridges with a concealed shingle-over type ridge vent. We vented the framed rake overhang by replacing one course of the yellow-pine wainscoting used as soffiting material with a strip of aluminum soffit vent.

The pierced and sawn vergeboards (see the photo on p. 29) are the dominant features of the exterior. We make the vergeboards from clear kiln-dried redwood 2x12 because any knots or checks would likely cause the delicate, short-grained pendants to break off. We laid out the design using a single-repeat template traced from the existing house, adjusting the spacing to get an even number of pendants. Sawing them out was a chore, even with a heavy-duty jigsaw.

Instead of a finial, the vergeboards meet at a simple square shaft, turned catty-corner and suspended from the peak. I wanted to go wild with an ornate spire, but the architects held me back. Some things haven't changed in 140 years. □

Scott McBride is a builder in Sperryville, Va., and a contributing editor of Fine Homebuilding. *Photos by the author except where noted.*

Updating the Rafter Square

A new scale quickly converts decimals to fractions, and a new table speeds the layout of hips and valleys

by Donald E. Zepp

Back in 1819, Silas Hawes, a blacksmith in South Shaftsbury, Vermont, became the first to receive a U. S. patent for a carpenter's square. Designed for basic layout work, the metal square had a long blade joined at the heel to a shorter perpendicular tongue. The blade and the tongue were engraved with inch scales, but little else is known about the square (records were destroyed in the Patent Office fire of 1836).

Over the next 70 years the square evolved into the more complicated version known today as the rafter square. Made of steel or aluminum, it has a 2-in. wide by 24-in. long blade and a 1½-in. wide by 16-in. long tongue. To help solve various layout problems, the edges of the square are graduated in either eighths, tenths, twelfths or sixteenths of an inch, and the sides are inscribed with math tables and scales.

Though building materials and techniques have changed dramatically in the last 100 years, the graphics on the rafter square have not. The face of the square (the side with the manufacturer's name stamped on it) typically has assorted rafter tables and a scale for laying out octagonal posts. The back has a gauge for converting fractions to decimals, a table for calculating board feet and a second table for calculating the lengths of diagonal bracing (for more on the rafter square, see "Tools for Building," pp. 30-35, from Fine Homebuilding's *Builder's Library* series, The Taunton Press, Inc.).

Some of these amenities are time-tested, others are timeworn. The octagon scale, for instance, is designed for laying out timbers having cross-sectional dimensions in full inches only (in other words, you can't lay out a 3½-in. by 3½-in. timber with it). This was okay 50 years ago, but not anymore. Also, the two rafter tables that are designed to lay out side cuts (bevel cuts) for hip, valley and jack rafters have been of limited value since the advent of the portable circular saw. The layout consists of a line scribed across the rafter's top edge at an angle that varies with the roof pitch and is indispensible (along with a connecting plumb line on the rafter face) for guiding side cuts made with a handsaw. Nowadays, however, side cuts are typically made by adjusting the shoe on a circular saw to 45° and cutting along the plumb line, rendering the side-cut layout line unnecessary. Side cuts are laid out only when the use of a circular saw is impractical, such as for making acute cuts on irregular-pitch roofs or for cutting thick rafter stock.

I've had plenty of experience using rafter squares, first as a carpenter during the 50s and 60s and then during 24 years of teaching advanced roof framing at the Williamson Free School of Mechanical Trades in Media, Pennsylvania. Early in my teaching career, I had a brainstorm: Why not replace the useless graphics on the square with helpful ones? I developed a scale and a table that, glued onto the square, instantly improve its performance (see sidebar facing page). In fact, one leading tool manufacturer seriously considered adding both to its squares (this is now on hold indefinitely because of high production costs).

The decimal-foot conversion scale—The length of any rafter can be calculated by multiplying the unit length of the rafter (listed under the appropriate rafter table) by the total run of the rafter in feet. If the total run is not in whole feet, or worse, is not in whole inches, you're stuck with having to do a bunch of risky, time-consuming conversions. One common method, for example, is to convert the total run to inches, multiply this figure by the unit length, divide by 12 to get an answer in inches, divide by 12 again to get an answer in feet and then convert, through another series of tedious calculations, any resulting decimal to inches and fractions of an inch.

You can avoid this drudgery by buying—at a cost of up to $70—a feet-inch calculator, which easily handles calculations involving mixed units of measure. Back when I was a carpenter, though, I carried a simple scale in my wallet for converting inches and fractions of an inch into hundredths of a foot. This budget feet-inch calculator was not only a big help in solving roof-framing problems, but it was a boon for solving *all* problems on the job site that involve feet, inches and fractions of an inch.

At first I taught each of my students to make one of these scales and to fasten it to their toolbox. Later I had them glue the scales to their rafter squares right over the octagon scale (see tongue of square, facing page). Here's an example of how the conversion scale works.

Say you're about to frame a gable roof with an 8-in-12 pitch, and you need to calculate the lengths of the common rafters, which have a total run of 13 ft. 5¼ in. First, use the decimal-foot table to convert 5¼ in. to .44 ft. (top photo below). Then look up the unit length of the rafters (14.42) in the appropriate rafter table. Multiply 14.42 times 13.44 to get 193.8 in., then divide that by 12 to get 16.15 ft. Finally, use the decimal-foot table to convert 16.15 ft. to 16 ft. $1\frac{13}{16}$ in. (bottom photo left) The problem was solved using just two simple calculations. (Don't forget that the result doesn't account for ridge thickness or rafter tails.)

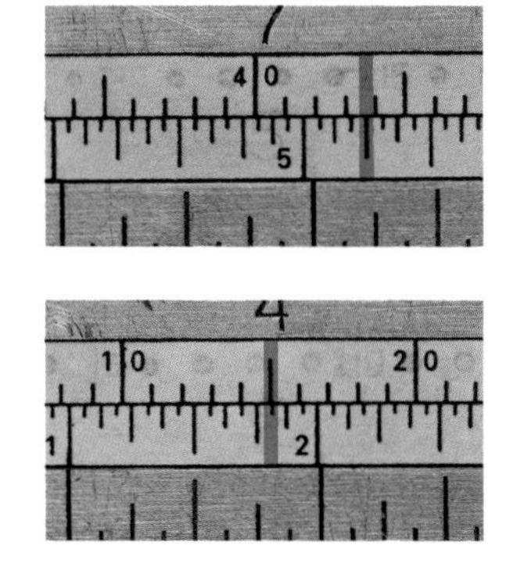

Jack-placement tables—One way to figure the lengths of the hip or valley jack rafters for a roof of a given pitch and rafter spacing is to consult the tables labeled "difference in lengths of jacks" on the rafter square. These give the unadjusted

Copy these

Here are full-scale drawings of a decimal-foot scale (top two strips) and a "difference in placement" table (bottom four strips), which can be photocopied and glued to your rafter square. For durability, the author bedded his scale and table in polyurethane varnish, let the varnish dry and then applied a second coat over the top.

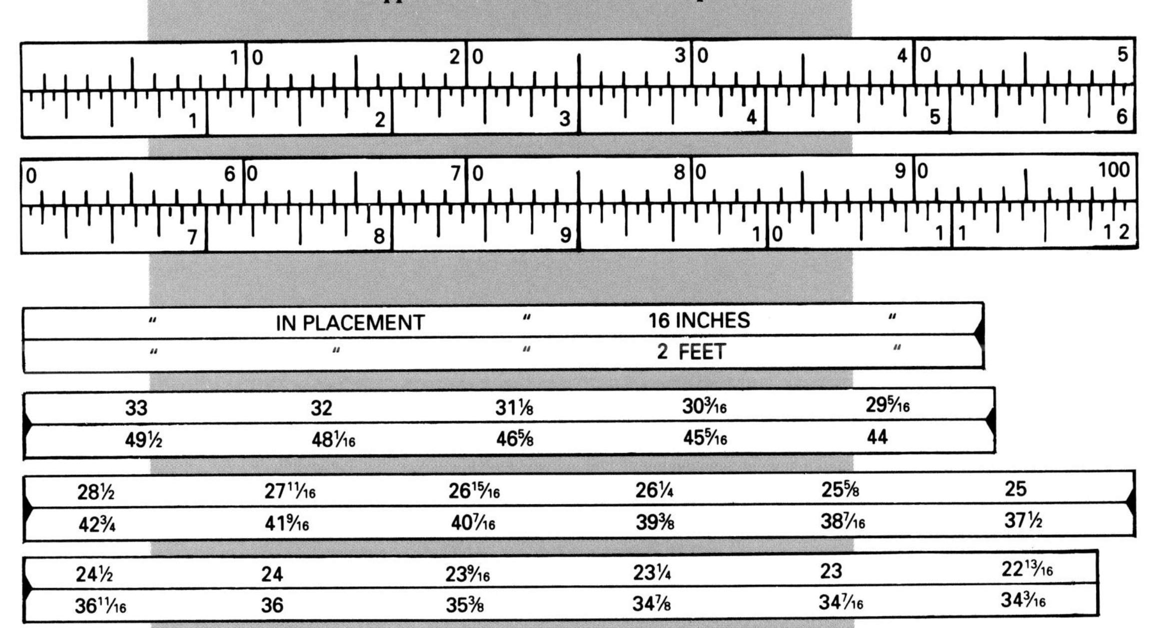

"	IN PLACEMENT	"	16 INCHES	"
"	"	"	2 FEET	"

33	32	31⅛	30 3/16	29 5/16
49½	48 1/16	46⅝	45 5/16	44

28½	27 11/16	26 15/16	26¼	25⅝	25
42¾	41 9/16	40 7/16	39⅜	38 7/16	37½

24½	24	23 9/16	23¼	23	22 13/16
36 11/16	36	35⅜	34⅞	34 7/16	34 3/16

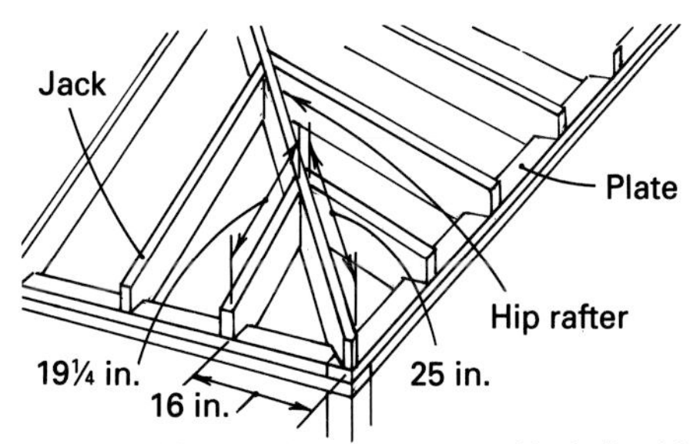

The "difference in placement of jacks" table (photo left) provides the on-center spacing of jack rafters along hip and valley rafters

lengths of the shortest jacks. The rest of the jacks in the roof are multiples of these lengths.

If the roof is layed out, cut and assembled accurately, and if the rafter stock is straight, there's no need to mark where the jacks should fit against the hip and valley rafters because the jacks will automatically land in the right place. In fact, the spacing will almost always vary. When roofs were decked with 1xs, this didn't matter. But now that plywood decking is the norm, variation in the spacing of jack rafters can mean that the plywood edge won't land on a rafter.

That's why I developed the "difference in placement of jacks" tables. Glued over the rafter side-cut tables (photo top of page), the tables tell where the jack rafters should meet the hips and valleys of the roof.

Say, for example, that you're framing an 8-in-12 hip roof with the rafters spaced 16 in. o. c. The "difference in length" tables will tell you that the jacks should be multiples of 19¼ in., and the "difference in placement" tables will tell you that the jacks should be laid out 25 in. o. c. along the hip rafters (drawing and photo above). □

Donald E. Zepp taught carpentry for 24 years at the Williamson Free School of Mechanical Trades in Media, Pa. He now lives in Dexter, Me.

Drawings: Christopher Clapp

From *Fine Homebuilding* magazine (June 1992) 75:50-51

All Trussed Up

Two structural systems that blend aesthetics and engineering

Steel-And-Wood Trusses

by Eliot Goldstein

We were at a critical point in our design of the Siegel house. All the rooms seemed to be in the right places, but we still hadn't been able to resolve the design of the entry hall. The essential components of the space—a grand staircase, high ceilings and lots of daylight—had been identified, but there was still some question about how to put it all together. We needed a dramatic element to bring the space to life. And we also had to address the structural forces at work on the entry-hall roof.

The two perpendicular wings of the house would meet at the entry hall, its steep gabled roof running from the front door to the rotunda-like family room. The outward thrust of the gable rafters (they would span more than 16 ft.), would have to be resisted somehow. Though the structure throughout the rest of the house would be hidden from view, I thought exposed trusses and a cathedral ceiling seemed to be the answer here. I decided on a truss system to resist the outward thrust of the gable rafters, and my codesigner on this house (and wife), Risa Perlmutter, agreed that such a system would keep the entry hall bright and open. Trusses would also be relatively economical and could be assembled on site by the framing carpenters. Finally, a truss system would present the opportunity to create the dramatic element we'd been seeking (photo left).

Design and engineering—My initial sketches of the truss showed all of its members in wood. Our structural engineer, Nandor Szayer, suggested that we consider instead using wood for compression members and steel for those in tension. After my clients endorsed the idea, he began to work out dimensions and connections (drawing facing page). We wanted to keep the truss members as delicate as possible, so it was important that they carry only axial (or longitudinal) loads. Unfortunately,

The entry hall. **The author wanted to create a sense of drama in the entry hall and still needed to resolve the structural elements of its roof. Addressing both issues resulted in separate rafter and truss systems, with the cedar ceiling seeming to float unsupported.**

Photo: Scott Frances/Esto

From *Fine Homebuilding* magazine (December 1991) 71:74-77

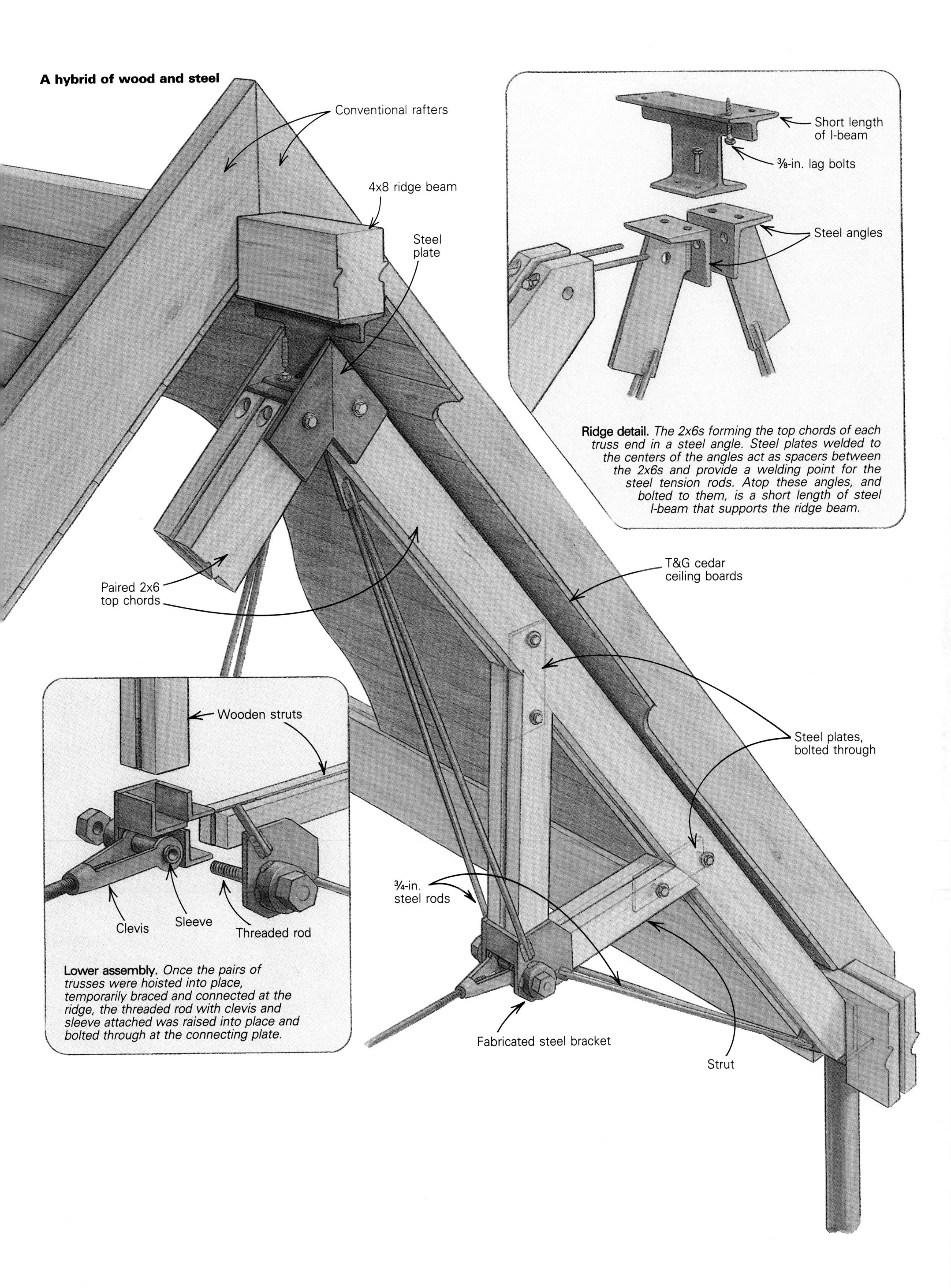

Ridge detail. *The 2x6s forming the top chords of each truss end in a steel angle. Steel plates welded to the centers of the angles act as spacers between the 2x6s and provide a welding point for the steel tension rods. Atop these angles, and bolted to them, is a short length of steel I-beam that supports the ridge beam.*

Lower assembly. *Once the pairs of trusses were hoisted into place, temporarily braced and connected at the ridge, the threaded rod with clevis and sleeve attached was raised into place and bolted through at the connecting plate.*

Drawing: Bob LaPointe

top chords of roof trusses are usually uniformly loaded along their entire lengths, inducing combined loading (compression *and* bending), which requires greater member depths.

To avoid combined loading, we chose to have the truss support only the ridge beam. The rafters, in turn, would bear on that beam, and on another parallel beam at their lower ends, but not on the trusses themselves. We decided to leave a substantial gap between the bottom edges of the rafters and the top edges of the trusses so the cedar ceiling boards could be installed between them later on.

It's easy to overlook the fact that the stresses in a truss vary constantly, and that under certain circumstances (in heavy winds or under seriously unbalanced snow loads, for example), there may even be stress reversals (from tension to compression or vice versa). Consequently, Nandor had to analyze the individual trusses for various loading conditions.

We also wanted to ensure adequate lateral bracing parallel to the plane of the truss system. We sheathed the south gable wall with plywood to create a shear wall, and tied the lower ends of the rafters above the entry hall into the plywood-sheathed attic floor (of the house's two converging wings) to prevent distortion of the gable.

Making connections—We resolved to use *concentric* joinery as much as possible, as *eccentric* connections (those in which the force lines do not converge to a point) cause twisting of the joints, and ultimately, bending in the truss members. Also, it was as important to us aesthetically that our connection details be spare as it was to keep the truss members delicate. Attaining elegant, concentric joints proved a challenge.

Nandor proposed that the trusses be assembled as a pair of trussed-rafters (each of which, though only half of the completed truss, functions as a truss itself), and that they be tied together with a horizontal rod (the bottom chord), once in place. Five members would converge at each of the joints held by this rod. The conventional way of connecting many converging members is to extend a plate or other steel fabrication out from the point of convergence, in the plane of the truss, so that the

Fabricating Box-Beam Trusses

by Edwin Oribin

The first 25 years of my professional life were spent designing buildings in the tropical region of coastal northern Australia. Then, several years ago, with our children all grown up and on their own, my wife and I moved 1,500 miles south—away from the equator—to a comparatively colder and dryer climate.

Taking these climatic factors into consideration, I knew I'd have to incorporate insulation and passive-solar principles in the design of our new house. Also, as I'd be working on the house mostly by myself, I wanted the structure to be reasonably manageable. Of particular concern to me was the design for the roof system. I decided on plywood box-beam trusses because they'd be lightweight, strong and fairly economical. This type of truss would also permit me a more sculptural design.

The main consideration in designing the clamping jig for my roof trusses was the short glue-setting time summer allows. According to the manufacturer, at 86° glue should take 45 minutes to set. In fact, when I tested it, it turned out to be more like 35 minutes. I was pretty certain that even with the simplest, quickest jig I could devise, we wouldn't be able to glue, assemble and clamp down the entire truss half in less than 35 minutes. I decided to do it in two stages instead.

Building the jig was a straightforward affair. I marked out the shape of the truss half on a temporary particleboard table, then edge-glued and screwed short 2x3 blocks around the perimeter and opposite each upright to define the form (drawing above). Next, I drilled holes adjacent to these blocks and fastened a ⅜-in. dia. threaded rod above and below the table. I left the rod projecting far enough above the blocks for nut, washer and another short piece of 2x3 (with holes drilled at each end) to act as the top member of the clamp.

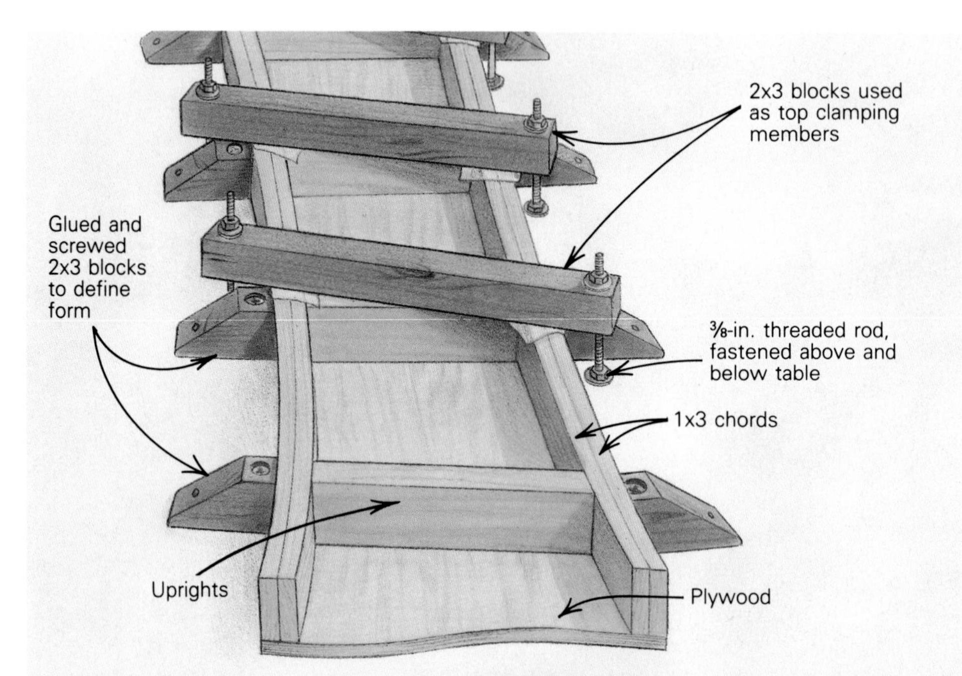

A site-built jig. *Using two sheets of particleboard end to end as a jig table, the author edge-glued and screwed small blocks of 2x3 to the table opposite each upright member to outline the form of the truss halves. Longer blocks of 2x3 span the width of the truss half and are bolted down through the table, forming the top member of each clamp. Each truss half required a two-stage glue-up because of the glue's extremely quick setting time in the summer heat.*

Preparation—Cutting to length the 1x3 top and bottom chords of the trusses went quickly. Then came the big job of cutting all the uprights. I took angles and lengths directly off of the jig, cut every piece on the radial-arm saw and numbered and stacked the pieces of each truss in a separate pile. I used a bandsaw to cut the irregularly shaped blocks that sit on the posts (drawing facing page). Our house needed 12 trusses—all told, I cut 408 pieces of timber and 96 sections of plywood (two pieces were required for each face of each half). I cut the plywood slightly undersize so that each section would drop smoothly into place between the blocks of the jig.

Having decided to assemble the truss halves in two stages, I still wanted to do a time trial. Working with my wife, we did a simulated dry run, complete with alarm clock. The simulated "glue-up" took us about 30 minutes. Feeling confident that we could duplicate the effort, we prepared to begin the real glue-up.

The urea formaldehyde glue powder we used had first to be mixed with water. It's hard to get all of the lumps thoroughly dissolved, but a rotary egg-beater did the trick. Fortunately, setting time doesn't begin until the hardener is added, but at that point there's only time for a quick final stir—then you've got to move fast.

Assembly—The first stage in our assembly procedure was to set the bottom sheets of plywood in the jig, then the pairs of 1x3s for the top and bottom chords, followed by all the uprights. All were glued as they went in, then clamped down (drawing above). I spread wax paper between the truss uprights and the 2x3 clamping blocks to keep the truss from be-

Drawings: Bob LaPointe

actual structural connections don't interfere with one another. This approach, while structurally sound, didn't strike us as particularly interesting or expressive.

We chose an alternate solution, one which involved layering the connection. Each member was assigned a separate layer within the thickness of the truss, thus allowing all members to converge at, or very near, a single point (as seen in elevation). Each joint consists of two pairs of wood struts (one pair horizontal, one pair vertical) housed in a fabricated steel bracket, with two pairs of diagonal rods welded to the outside faces of the bracket. A Y-shaped clevis is bolted to the bracket, and a horizontal tie rod threads into the other end of the clevis. The two ends of the tie rod are threaded in opposite directions to allow tensioning (drawing, p. 37).

Assembly and installation—To ensure consistent results, the fabricator welded most of the steel components together in his shop and primed them, which would prevent corrosion while the members were exposed to the weather. On site, the carpenters simply cut and drilled the 2x lumber, then bolted it to the steel.

The steel columns on which the trusses sit were already in place. The trussed-rafters were lifted up and bolted to the cap plates of the columns; temporary bracing was installed to keep the trusses from falling like dominoes. After the tie rods had been adjusted to the proper length, the 4x8 wood ridge beam was installed. Conventional rafters were then birds'-mouthed onto the ridge and eave beams, then sheathed with plywood. A continuous ridge vent was installed.

The finished ceiling boards were not installed until the house was tight to the weather. The carpenters had left a larger gap than we'd planned between the undersides of the rafters and the tops of the trusses, but we actually like it better. The visual separation of the primary structure (the trusses) from the secondary structure (the rafters) parallels their functional distinctions. □

Eliot Goldstein is an alpine design consultant and a partner in the firm of James Goldstein & Partners, Architects, in Millburn, N. J.

The house completed. **The author wanted a simple, economical, roof support system that had character. The sculptural flair of opposing curves is evident here.**

coming glued to the jig. When the glue had set we undid the clamps, glued and fitted the top sections of plywood and resecured the clamps.

Once out of the jig, each truss was finished around the edges with a router set to cut the plywood ¼ in. inside the line of the timber. That gave the trusses a nice rabbeted edge. I applied one coat of stain/sealer, fitted the purlin support blocks and stacked the trusses on edge with spacer blocks between.

I would hate to pay for trusses like this made in a factory, but the cost of the materials was very reasonable, and I had the time. In my situation, it was a very economical way of spanning 20 feet. □

Edwin Oribin is an architect in Stanthorpe, Queensland, Australia.

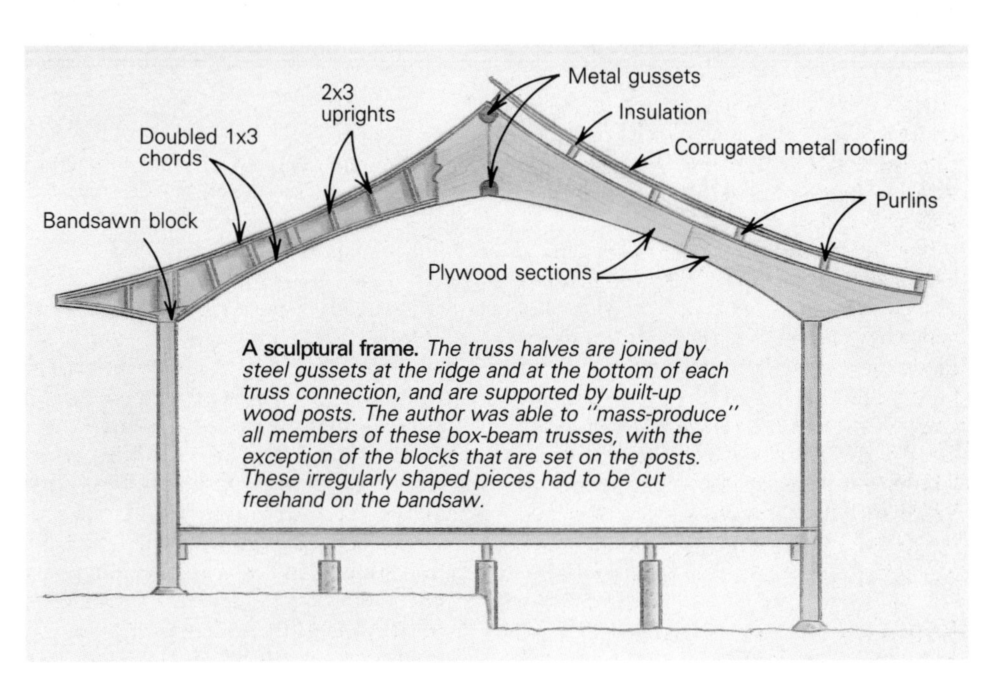

A sculptural frame. *The truss halves are joined by steel gussets at the ridge and at the bottom of each truss connection, and are supported by built-up wood posts. The author was able to "mass-produce" all members of these box-beam trusses, with the exception of the blocks that are set on the posts. These irregularly shaped pieces had to be cut freehand on the bandsaw.*

Examining Carbide-Tipped Sawblades

A look at blade design and manufacturing will help you choose the right blade for the job

by Sanford Wilk

On a recent trip to the local tool outlet, I overheard a customer ask a young salesman what blade was needed for ripping stock on a table saw. The enthused salesman recommended a 40-tooth blade with "alternate top bevel" (ATB) grind, and the customer blithely checked the price and bought the blade.

Within minutes, the front-door chime sounded, and a large man with muscles on his ear lobes walked through the door. He exchanged greetings with the salesman and said he was looking for a framing blade to use on his "granddaddy of a worm drive." The salesman handed the framer a 40-tooth ATB, the same kind that he had just sold the previous customer, and without even a second look, the big man grunted, "Put it on my account."

At that point I was getting a little suspicious, so I asked the kid what blade he'd recommend for use on my slide/compound-miter saw. "I want a blade that's good for both finish and rough work and that won't kick back on me." Sure enough, he handed me a 40-tooth ATB blade.

The saddest part about the whole scenario wasn't that the salesman was poorly informed, but that like the two guys before me, I'd never given much thought to sawblades. That had to change.

I started reading up on sawblades, talked to various tradesmen about them and interviewed (and even visited) some manufacturers. I'll tell you what I learned about how sawblades are designed and manufactured, which should help you make informed choices when buying and using blades.

Unfortunately, not all the pertinent information about a blade is on its package. A reputable dealer, however, should be able to provide additional information about a blade. You can also call the manufacturers directly. In some cases, you may not be able to find out certain things—the Rockwell hardness of a blade, for instance. But even knowing what variables affect blade quality and cost helps you to make an informed choice.

This article will deal generally with 6½-in. to 10-in. blades, those used on portable circular saws, crosscutting saws (miter, radial-arm, slide/compound) and table saws. I've limited the discussion to carbide-tipped sawblades because, in my experience, most professionals don't even bother with plain steel blades anymore. Why? Well, besides the fact that carbide-tipped blades will last up to 60 times longer than steel blades, the quality of the cut is significantly better. Blade selection depends not only on price range, but also on what material you're cutting and on what saw you're using. The chart on p. 45 lists some guidelines for blade selection.

From left to right on the facing page: Freud's 40-tooth finishing blade with antikickback shoulders; Vermont American's 24-tooth Lightning crosscut blade; and SystiMatic's 24-tooth ripping blade.

Sawblade trade-offs—Sawblades are made in various grades (and price ranges) to meet the needs of three general markets: industrial users, such as cabinet factories, where saws are running constantly; professional woodworkers, including carpenters and cabinetmakers; and do-it-yourselfers (DIY) involved in home maintenance or amateur woodworking. As you move from industrial blades to DIY blades, manufacturers reduce costs (and quality) by using cheaper materials and shortcutting the manufacturing process. But depending on the application, a builder might reasonably select a blade from any price range, so it's helpful to know specifically what features account for one blade costing $5 and another, very similar looking blade, costing $50.

These days, a decent 7¼-in. blade costs between $17 and $26. The same quality blade for a 10-in. table saw will run about four times that figure. If your saw's arbor alignment is off (due to such problems as excessive bearing wear), then even the best blade isn't going to produce a clean cut. Nice equipment deserves a quality blade and vice versa.

Materials and manufacturing aren't the only areas of compromise, however, because blade design represents another series of trade-offs. Performance is traded for versatility. Ripping teeth, for instance, can be combined with crosscut teeth on the same blade, resulting in a general-purpose blade that neither rips nor crosscuts as well as a specialized blade. Another example is that speed is traded for smooth cuts by adding more teeth to a blade.

Blade blanks—The blank or body of the blade plays a major role in overall blade quality. Blade blanks are generally either stamped out on a punch press or cut by a laser. A stamping machine with a dull cutting die can produce cupped blanks. The laser-production technique is less prone to distorting the blank.

After the blank is cut, it is tempered; that's where hardness comes into play. The blank's steel is rated in "Rockwell hardness" (Rc) the industry standard unit of measurement. Most manufacturers don't disclose their blade's Rc rating on the packaging, but a reputable dealer should be able to give you the information.

I prefer a hardness of at least 44Rc, but good blades can be found in the hardness range of 35Rc to 48Rc. Beyond about 50Rc, the steel becomes more brittle, and the blade more susceptible to cracks. In general, thin-kerf blades need a higher Rc rating than standard blades because without harder steel, they would bend from side forces due to twisting or binding of the workpiece. I wouldn't buy a thin-kerf blade with an Rc rating of less than 40.

If you get a blank in the 40Rc-or-better range, or a blade in the 35Rc-or-better range that has been properly tensioned (more on this later), then that blade will be less vulnerable to distortion and twist and more likely to remain flat. Keep in mind that if a blade wobbles (called *lateral run-out*) because it's distorted, blade life will be shortened by the excess strain, your saw can be damaged and the tolerances of your work will suffer. Here, too, it's best to follow the old cliché of "better safe than sorry." Stick to a suitable steel-hardness rating, and you'll have one less thing to worry about.

After a blade blank has been cut, it is ground (or sanded) flat. If you look at a cheap blade, you'll notice that the grind marks look coarse or spiral outward (rather than being concentric). Spiral grind marks indicate that the blade was sanded, which is not as precise a process as grinding. Poorly ground (or unground) blades suffer from distortion and burrs, and little things like that can cause a blade to wobble.

Next (with some blades) comes a process called roll-tensioning. This compresses the blade across its thickness to introduce internal stresses that compensate for centrifugal forces and for heat expansion that would otherwise distort the blade when it's being used.

Originally, blades were flattened and tensioned by highly skilled craftsmen called sawsmiths. It's hard to imagine, but these guys banged at blades with hammers all day long until each and every blade was true and properly tensioned. There may still be some sawsmiths around (especially in Europe), but today most blades are tensioned by machine.

Roll-tensioning is done with a machine that positions one wheel on top of a blade blank and another on the bottom. The wheels squeeze the blade to compress it and force the outer rim into compression. Roll-tensioning does not cure flatness problems, however, just as a thickness planer won't flatten a cupped board.

Many blades that are 10-in. or less in diameter aren't tensioned at all in the United States. In Europe blade makers will tension smaller diameter blades, but most of their American counterparts feel differently about this and leave tensioning to larger blades. Their argument is that only bigger blades suffer from high rotational stresses. The bottom line is that a tensioned blade will run truer, cut smoother, last longer and cost more than a blade that isn't tensioned.

The arbor hole is also a measure of a blank's quality. On top-notch blades, the arbor holes are reamed or ground to precise size after the blanks are made. Arbor-hole size is critical because oversized or elongated holes can allow high peripheral run-out. Instead of cutting in a perfect circle, the blade tends to cut in an ellipse pattern, which means a rougher cut, more wear on the blade and, hence, more frequent sharpening.

Most of us don't take our saws into the store when we select blades, so it's a good idea to check the store's refund policy before you buy. If the dealer won't let you return a blade that doesn't fit your arbor snugly, then you might be better off buying elsewhere.

Thin-kerf blades—Our obsession with decreasing circular-saw weight has led to a decrease

From *Fine Homebuilding* magazine (February 1992) 72:42-47

in the available power of today's saws. This is one of the reasons why thin-kerf sawblades have become so popular. The difference in kerf size offered by this slimmed-downed blade style seems to be about 1/32 in. (or about a 25% reduction) over conventional blades (photo right). That savings results in less stock removal, lower friction and—you guessed it—higher feed rates, while using less horsepower. Good idea, eh?

Actually, thin kerfs aren't the cure-all, either. In exchange for a thinner tooth, you inherit a blade that distorts more readily (remember that the blade's body gets trimmed down along with the tooth). Unless your saw is too underpowered for a full-kerf blade, I don't recommend thin kerfs for most operations. These blades are popular because of the gains that I noted, but payback is typically lower because the blade sacrifices longevity and risks deflection, which can make a real mess of that project you've put so much effort into. One place where thin kerfs are a better choice is in rough framing, where minor blade-cut deflections are usually easier to forgive.

Blade thickness. **From left to right, the blades pictured above are heavy duty, thin kerf and ultra-thin kerf. Thinner blades require less power from the saw and less push from the operator. Unfortunately, thinner blades don't last as long as heavy-duty blades, and thin kerfs may be more prone to distortion.**

Detailing a blank—The space between the teeth of a sawblade is called the gullet (photo facing page) and serves to contain the sawdust produced by each tooth until it has a chance to resurface from the cut and expel the waste.

Many blades have four heat-expansion slots punched or laser-cut into their bodies. These slots extend from the base of a gullet toward the arbor and end in a small hole. Heat can cause a blade to distort (as well as causing resin to build up on the blade). Heat-expansion slots help the blade to counter the effect of heat buildup by allowing room for the steel to expand. Because their thinness makes them more prone to distortion anyway, many thin-kerf blades sport heat-expansion vents in addition to slots. These vents are isolated holes of various shapes located in the middle of the blade.

Some blades feature grossly oversized gullets, as well as expansion slots and vents in weird configurations. Some of these are just marketing gimmicks. Stick to simple expansion slots and to a gullet depth approximately equal to the depth of the tooth face.

The last features that you might want to consider when choosing the right blank are surface treatment and noise limiters. I feel that a silicone or Teflon coating helps the blade to cut through wet, icy or pressure-treated lumber. Some of the cheaper blades, however, are simply covered with Teflon paint, which can wear or flake off easily. Better blades are covered with sintered Teflon. This involves the use of heat and pressure to bond Teflon powder to a blade.

Keep in mind that a nonstick coating doesn't last forever, and that sooner or later the coating begins to wear, and the blade begins to gum up. When the blade begins to stick, it should be cleaned; a clean blade cuts better. Pizza pans work well for soaking sawblades. For a solvent, you can use kerosene, alcohol, ammonia or even oven cleaner.

Blade noise is caused by blade vibration and by air passing through the heat-expansion slots. Some blades have plugs inserted into the end of the slot as a noise-damping device. The plug, usually brass or copper, is a softer material than the blade, so it absorbs most of the vibration and therefore the sound. New laser technologies allow heat-expansion slots to be cut more efficiently and to be smaller. Laser-cut slots are still effective for heat control but don't require plugs. Here again is another advantage to laser-cut blanks. In general, thinner blades with shallow gullets should cut quieter.

Carbide teeth—Carbide teeth are composed of small grains of tungsten carbide suspended in a cobalt binder. The size of the carbide grains and the relative percentage of the binder determine how hard (and abrasion-resistant) the tooth will be. Smaller grains of carbide make for a denser, stronger tooth (depending also on the percentage of binder used). The newest tooth designs employ smaller carbide particles (micrograin carbide) than ever before. The result is a densely packed carbide tooth with exceptional cutting power and longer life.

Most circular sawblades make use of carbide rated from C-1 to C-4. (To give you some idea of how hard this stuff is, C-1 carbide is about 89Rc and C-4 carbide is about 94Rc.) But don't assume that hardest always means best. It doesn't. For applications in which you are likely to hit a nail (such as cutting through a roof), a softer grade of carbide is usually better because it is less likely to chip. Harder carbides, like C-3 and C-4, are helpful when you're cutting laminated plastics or particleboard—materials that can dull a softer carbide tooth quickly. Using the right carbide for the job is important because damaged teeth are usually difficult and expensive to repair.

Once the tooth is formed, it is brazed or welded to the blade blank. If you look closely (use a magnifying glass), the brazing will either appear to be consistent and homogenous, or it will look pitted. Pits are usually a sign that the actual braze is weak and that the blade might lose a tooth. The loss of a tooth at 5,800 rpm isn't just scary—it's dangerous. The possible loss of an eye isn't worth the few bucks saved by buying a cheap blade. But tooth loss is possible even with the best of blades, and I take wearing safety glasses as seriously as I do buckling my truck's safety belt. Consider what kind of craftsman you would be without your sight....

In the Black & Decker blade lab, researchers test various makes of blades to find out what the competition is up to. When I visited this facility last summer, the lab people were running a test that made repetitive cuts without a breather. In the limited time that I was there, I noticed that almost all of the inexpensive carbide blades that were tested dulled quickly.

One of the ways that blade manufacturers cut costs is by making the tooth smaller and thinner. The relatively smaller tooth of a cheap blade will wear away after just a few sharpenings. The eventual savings to the craftsman aren't that great when you consider that these less-expensive blades typically only go half the distance on resharpenings.

On the flip side, some blades have larger teeth to permit extra sharpenings and thus longer life. But larger teeth can place an added burden on saw motors that are usually overtaxed to begin with.

Carbide-tipped blades cannot be sharpened properly without special equipment, such as diamond grinding wheels. Even then, the direction of the grinding and the tooth's finish are critical to restoring the blade's intended edge. The point here is that for most of us, it's clearly less expensive and more convenient to have a reliable sharpening shop put the edge back on our investment rather than to try sharpening the blades ourselves.

Tooth configurations—There are four basic tooth configurations, or grinds, available. The first is the *flat-top* (FT) grind (also called

"square top," "rip grind" or "chisel tooth"), which looks just as its name implies (photo A, next page). It cuts a kerf with a smooth, square bottom, employs a limited number of teeth and is designed for ripping. Because the blade's teeth are set at an aggressive hook angle (20° or more), this blade's nature is to gouge the stock. (The hook angle is the angle the tooth makes with the line of the blade's radius.) The steep hook angle combined with a large gullet translates to faster stock removal and speedier cutting, but you will almost certainly splinter the grain if you use the blade to crosscut. Keep in mind that it's easier to cut with the grain than against it.

The next tooth configuration is the *triple-chip grind* (TCG). Here, every other tooth is a flat-top tooth, called a raker, followed by a tooth that has been chamfered on each side (photo B, next page). The chamfered tooth chips out the center of the kerf, leaving the uncut edges for the raker to clean up. The "triple chip" name comes from that fact that the chamfered tooth removes one chip (from the center of the kerf) and the raker removes two more (one on either side of the kerf). The TCG design works well with particleboard, laminates, hardwoods and nonferrous metals. I like this blade design for cutting synthetic countertop materials (like Corian and Fountainhead) because it is less prone to wear.

The third tooth configuration is the *alternate-top-bevel grind* (ATB). Here, the tops of the teeth are ground at an angle, with alternate teeth ground in alternate directions (photo C, next page). The result is that the lead portion of each tooth is a point rather than an edge, and these points sever the grain rather than tear it. Hence, this tooth is designed for crosscutting and will rarely splinter the stock. Because it lacks a raker, however, the bottom of the kerf resembles a "V" in shape. That isn't a big deal unless you're using the saw to provide decorative cuts that require flat-bottomed kerfs.

The final basic tooth configuration is the *alternate top bevel with raker* (ATB&R). This blade type seems to be the most popular design. For every four ATB teeth, this blade employs one raker tooth that cleans out the kerf (photo D, next page). Each set of five teeth is separated by a fairly deep gullet. An ATB&R blade has a total number of teeth that is divisi-

A. Flat-top grind. Designed for ripping, this blade has teeth whose tops and faces are ground square, have deep gullets and steep hook angles.

B. Triple-chip grind. On this blade, every other tooth is ground square, followed by a tooth that's beveled on three sides.

C. Alternate-top-bevel grind. Alternate teeth are ground to a point in opposite directions so they sever the grain when crosscutting.

D. Alternate-top-bevel with raker. For every four teeth with alternate top bevels, this blade has one square tooth called a raker.

ble by five. I use ATB&R blades on my table saw because they cut cleanly and can be pushed hard. This type of blade is a jack-of-all-trades (and thus a master of none). But it does handle most materials (including plywoods and particleboards) reasonably well. Thanks to the tooth and gullet design, you can get a fairly quick rip or crosscut, but you sacrifice a clean cut.

Blade manufacturers are always amending these standard tooth types and creating their own special hybrids that evolve from the standard designs. These new designs may be less noisy, more efficient and cleaner cutting, but some sharpening shops don't have the equipment to put a proper edge back on the teeth. Before buying such a blade, it would be wise to check and see if a local shop will sharpen a unique tooth design. You don't want to get stuck with an expensive "throwaway" blade.

Tooth type and blade performance—Ever wondered how many teeth a blade should have? It depends. In general, if you want extremely clean cuts, go with a lot of teeth. But keep in mind that more teeth mean a slower feed rate and higher blade temperatures, which translates to more resin gumming up the blade and more frequent sharpenings. If you don't care about the smoothness of the cut but would rather have lightning-quick speed (when framing, for instance, or when you plan to joint the stock anyway), then opt for fewer teeth.

This is an oversimplification, however. A better way of determining the optimum number of teeth on a blade involves the thickness of the stock you're cutting. For ripping, three to five teeth should be cutting (don't count the teeth that are exiting the cut). For crosscutting, five to seven teeth should be cutting. These rules assume that blade height (or depth) has been adjusted to extend ⅛ in. above the stock, which is not only the proper adjustment for safe operation, but also for smooth cutting.

Yet another consideration beyond those of tooth style and quantity is the hook, or rake, angle. An angle of 30° or greater results in a faster cut. As you might imagine, there's a trade-off here, too, because a high rake angle means more chipping and splintering.

Low hook angles (5° and 10°) are slower but offer a smoother cut. Probably the most interesting blade for finish carpenters is the "negative-rake" design. This blade actually has teeth that lean away from the blade's rotation rather than into it. For those of you experiencing a "pull" from your radial-arm saw or sliding compound miter, this blade spells relief. A blade with a negative-rake design will cut smoothly and slowly. But speed isn't really important when crosscutting on radial-arm saws or miter saws.

A recent development in blade safety is Feruds' new antikickback design, which features teeth with extremely high shoulders (called limitators) behind them (far left blade, photo p. 40). According to the folks at Freud, these high shoulders limit the bite of each tooth, which reduces the chances of a kickback and greatly reduces the force of any kickback that does occur. □

Sanford Wilk is an architect, builder and writer in Boston, Mass. Photos by Susan Kahn.

Rules of Thumb

As you move from industrial blades to professional blades to DIY blades, quality is traded for lower cost by using cheaper materials and shortcutting the manufacturing process.

An expensive blade will typically last longer than a cheap blade, and the cost difference is usually made up by the number of extra sharpenings available from your investment.

A blade that has been tensioned will run truer and cost more than a blade that hasn't been tensioned.

Blades with low hook angles (5° and 10°) cut more slowly but offer a smoother cut.

Carbide-tipped blades will last up to 60 times as long as steel blades.

Silicone or Teflon coating helps when cutting wet, icy or pressure-treated lumber.

A clean blade cuts better.

SAWBLADE RECOMMENDATIONS

	Rip cut	Crosscut	Pressure-treated wood	Plywood	Particleboard, plastic laminates	Synthetic/ solid-surface countertops
Portable circular saw	15°-25° hook 16-32 teeth FT/ATB	5°-20° hook 16-60 teeth FT/ATB	Rip or crosscut blade with Teflon or silicone coating	5°-20° hook 24-40 teeth FT/ATB	5°-20° hook 24-40 teeth FT/TCG	5°-20° hook 40 teeth FT/TCG
Table saw	15°-25° hook 20-40 teeth FT/ATB	5°-20° hook 40-80 teeth ATB	Rip or crosscut blade with Teflon or silicone coating	5°-15° hook 40-60 teeth ATB/TCG	–5°-15° hook 40-80 teeth TCG	–5°-15° hook 40-80 teeth TCG
Miter saw	not applicable	–5°-15° hook 40-80 teeth ATB	Rip or crosscut blade with Teflon or silicone coating	–5°-15° hook 40-80 teeth ATB/TCG	–5°-15° hook 40-80 teeth TCG	–5°-15° hook 40-80 teeth TCG
Slide/compound saw	not applicable	–5° hook 40-80 teeth ATB	Rip or crosscut blade with Teflon or silicone coating	–5° hook 40-80 teeth ATB/TCG	–5° hook 40-80 teeth TCG	–5° hook 40-80 teeth TCG
Radial-arm saw	12°-20° hook 20-40 teeth FT/ATB	–5°-15° hook 40-80 teeth ATB	Rip or crosscut blade with Teflon or silicone coating	–5°-15° hook 40-60 teeth ATB/TCG	–5°-15° hook 40-80 teeth TCG	–5°-15° hook 40-80 teeth TCG

FT = Flat-top grind
TCG = Triple-chip grind
ATB = Alternate-top-bevel grind
Hook = The angle the tooth makes with the line of the blade's radius

ATB&R (alternate-top-bevel with raker) blades are not included on the chart because they are combination blades designed for versatility, not for optimum performance. ATB&R blades will work for many of these applications but will not perform as well as the blades listed above.

The ragged edge. **A circular saw left a trail of destruction across the grain of this ½-in. AC plywood. This could easily have been prevented.**

Cutting Across the Grain

These simple techniques will help you to make cross-grain sawcuts that are smooth as silk

by Larry Haun

I remember the first door I was allowed to hang as an apprentice carpenter in the early 50s. We were doing finish work in a large custom home, and my task was to haul things around and nail in some of the pieces that the journeymen were cutting and fitting into place. They had installed a number of doors, and toward the end of one day, the foreman allowed me to hang a simple veneered door in a back room. I did great, except that I forgot to score the veneer before cutting the door to length. The splintered edge I left behind marred an otherwise commendable effort. Well, I learned the hard way, but nobody should have to. Cutting across the grain of a door or a sheet of plywood (photo above) without messing up the edge can be done successfully with minimal fuss. The simple solutions I'll describe here can also minimize problems when cutting other materials, including particleboard, Masonite and plastic laminates.

First, the sawblade—One solution couldn't be much simpler: Choose the right blade for the material you're cutting. The teeth of a sawblade tend to break slivers from the surface of the material being cut, and the wrong blade makes the problem worse, resulting in a ragged edge instead of a smooth one.

Carbide circular-saw blades with a triple-chip grind (TCG) and 40 teeth leave a clean cut on hardwoods, particleboard and plastic laminates. A sawblade with an alternate-top-bevel (ATB)

A guide to straight cuts. **This cutting guide includes built-in clamps that set easily to grip panel stock. The 50-in. length of the tool means it can guide cuts across the width of a plywood sheet.**

From *Fine Homebuilding* magazine (August 1992) 76:63-65

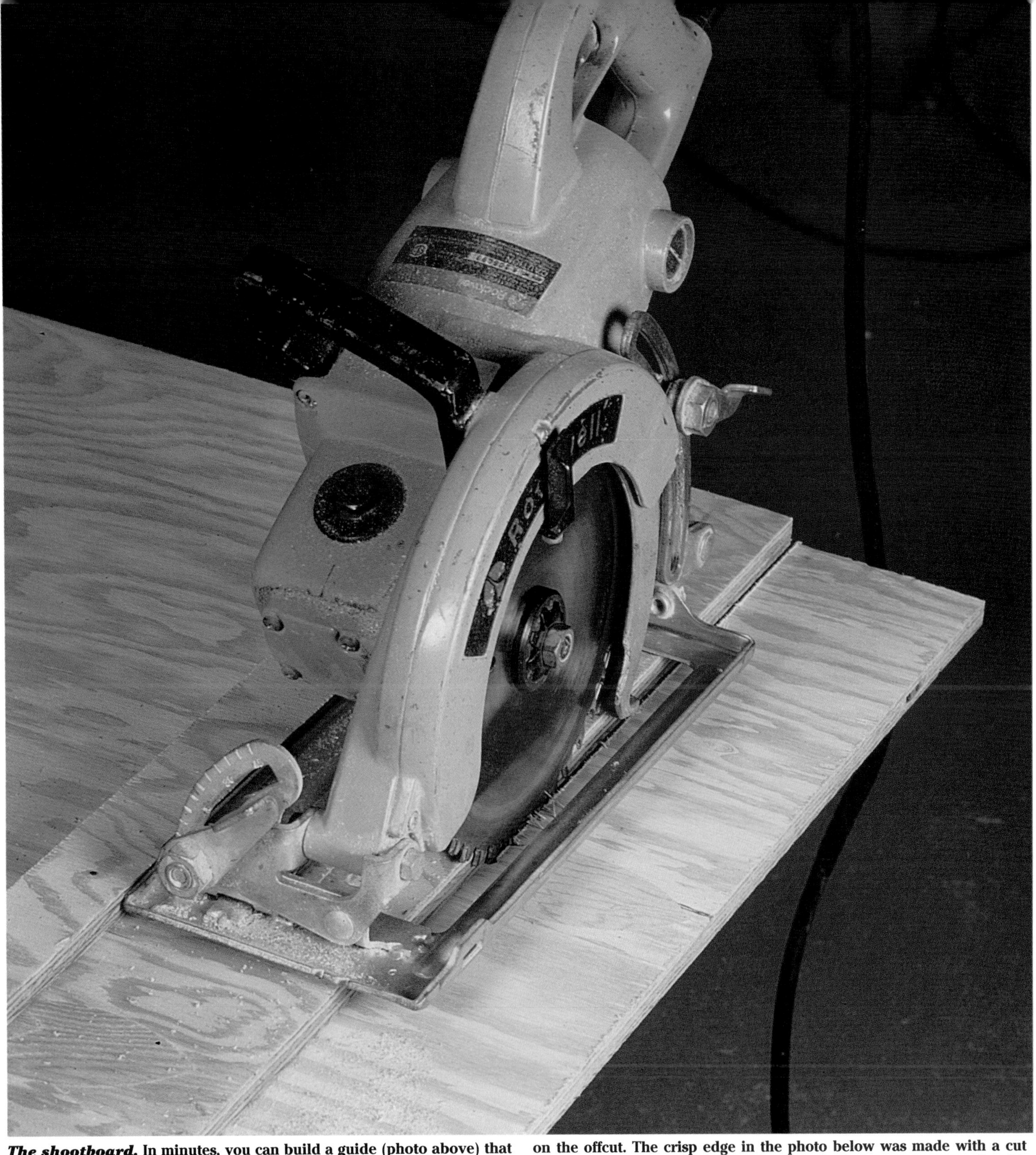

The shootboard. **In minutes, you can build a guide (photo above) that will ensure straight cuts and minimize cross-grain splintering. The guide is clamped to the "good" side of the workpiece. The splintering seen here is on the offcut. The crisp edge in the photo below was made with a cut guide. The cut shown in the top photo on the facing page was made with the same sawblade on the same material but *without* using the cut guide.**

grind—mine has 36 teeth—works well on plywood (see the article on pp. 40-45 for more information on TCG and ATB sawblades). Like any cutting tool, sawblades work best and leave a cleaner cut when they are sharp.

When cutting with any blade, especially when you want an absolutely straight cut, clamp a straightedge to the work to act as a saw guide. Edge chipping is minimized when the sawblade doesn't wobble in the cut. You can use all sorts of things as straightedges (as long as they're straight), but I use a store-bought 50-in. long aluminum edge that easily clamps to the stock. It's a simple and easy-to-use device that locks in place as you push a lever on one end, and the integral clamping mechanism adjusts to various widths (bottom photo, p. 46).

Manufacturers also make jigsaw blades that minimize tearout. Bosch, for example, offers one (#T-101BR) with a reverse-tooth configuration that leaves a chip-free upper surface when cutting plywood or plastic (photo below right). They also have a plastic insert that fits into the shoe of any Bosch jigsaw. The insert snugs up to the blade tightly and holds down veneers as they're being cut (Bosch Power Tools, 100 Bosch Blvd., Newbern, N. C. 28562; 919-636-4200).

The saw flange. **This homemade metal flange bolts to a circular saw's baseplate. The edge of the flange rides along the cutline to hold the wood fibers in place. Placed properly, it won't interfere with the operation of the blade guard.**

Making the score—The time-honored method of avoiding tearout is to lay a straightedge across the material and score the cutline with a sharp knife. This cuts through the outer surface of the wood fibers, preventing them from lifting up as the saw chews its way along the scrap side of the line. If you have only a few cuts to make, the technique works admirably. But anybody who has lots of doors to hang or stacks of cabinet-grade plywood to cut will find this method time-consuming.

The masking method—Another simple way to keep the chips down is to lay a strip of masking tape over the area to be cut, then mark the cutline on the tape. The tape holds the wood fibers in place during the cut and, once removed, will reveal a smooth edge. A couple of cautions, though. If you use tape with too much stick, you're likely to pull wood fibers from the stock as you pull up the tape. Cheap tape works just fine. Also, as you're removing the tape, pull it in a direction perpendicular to the cutline; this minimizes any disruption of the fibers. The tape technique works well if you have a few cuts to make, but it gobbles up time if you have a lot of them. By the way, the tape trick is also helpful if you work with textured plastic laminates or dark wood veneers. Mark your cutline on the tape, and it will be easier to see.

The shootboard—Yet another way to make a clean cut is with something we call a shootboard. It's simply a straightedge with a fence screwed or nailed to it (top photo, p. 47). The shootboard is clamped to a cutline marked on the material. The saw base rests on the straightedge surface and helps to hold down the cross grain while the fence guides the saw to a straight cut. Also, there's no marring of the finished surface by the saw base. Just remember to set the depth of cut deep enough to account for the additional material.

Shootboards are easy to make. Take a piece of ½-in. plywood or hardboard that's about 8 in. wide and secure a 1½-in. strip of the same material to one edge. Place your saw on the straightedge (and against the fence), then cut off any excess material. The edge of the shootboard will match the cutline precisely and will keep the wood fibers from lifting up (bottom photo, facing page). Commercial shootboards with an adjustable fence are available for around $30 (Olive Knot Products, P. O. Box 188, Corning, Calif. 96021; 800-759-6283 or 916-824-5280 in Calif.)

The saw flange—A device developed by California door hanger Al Schieffer some 30 years ago eliminates the need for all the techniques I've described, at least when you're using a portable circular saw. Schieffer modifies his saw by adding a homemade metal flange that bolts to the baseplate (photo above). The flange is bent so that it comes down alongside the sawblade and rides lightly on the material during a cut. The pressure of the flange prevents the wood fibers from lifting.

The great thing about this device is that it really works. The bad thing is that no one has developed it for the commercial market. I made mine from a 1-in. wide by ⅛-in. thick piece of metal plate strap in about an hour. First, bend the strap to fit the saw base, drill holes in the strap and then drill matching holes in the baseplate. Depending on the model of your saw, you might have to countersink the holes. It takes some adjusting to align the strap properly alongside the blade and also to be sure that it rides properly on the material being cut. But a bit of trial-and-error cutting gets the job done. Note: The edge that rides against the wood has to be rounded and smoothed with a file so that no scratches are left on the stock.

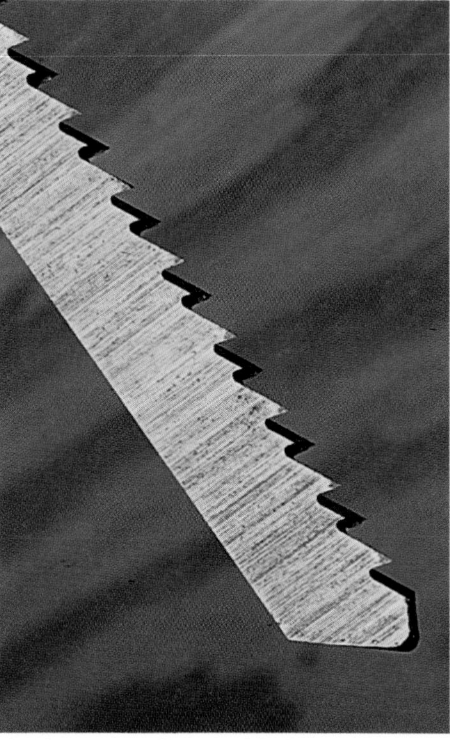

Anti-tearout jigsaw blade. **The reverse-tooth configuration of this Bosch blade leaves a chip-free surface behind.**

While you're remodeling your saw's baseplate, take one more step to ensure good cuts. I don't cut a lot of fine materials, but I did glue plastic laminate to the underside of the baseplate to keep it from scratching any finished surface. □

Larry Haun lives in Los Angeles, Calif., and is a member of Local 409. Photos by Robert Marsala.

Snapping a line. **A chalkline can be snapped across the tops of studs and cripples to mark a cutline. Before anything is cut to length, the framers will set the top plate on edge above the line and mark the framing layout on it. *Photo by Ron Turk.***

Building Rake Walls

Two time-saving layout methods

by Larry Haun

Most wall layout is quite simple. The process of transferring dimensions from prints to concrete slab or subfloor usually consists of little more than snapping a series of chalklines to form squares and rectangles. On occasion, however, plans will call for a room with a cathedral ceiling that follows the pitch of the roof. Here rafters double as joists, rising upward from an outside wall to the ridge. Gable-end walls in these rooms are called rake walls, and laying one out isn't much more difficult than laying out a regular wall. But being aware of a couple of simple techniques will speed up the process. The methods I discuss here have served me well for the past 30 years (for another approach to building rake walls, see the article on pp. 52-53).

The location of the bottom 2x plate of a rake wall is laid out on the floor like any other wall. The location of the rake wall's top plate is chalklined out at an angle from the top of the shortest stud. This way, the framer can build the wall without making any further calculations, even though each stud will be a different length. The angle of the top plate is determined by the pitch of the roof.

A calculated solution—There are two fairly easy ways of laying out rake walls. The first calls for a pocket calculator, which is used to determine the difference in length between the shortest and longest studs. The shortest stud is normally a standard length, 92¼ in., so once you've established the *difference* in length between shortest and longest, you know the *actual* length of the longest stud. With the heights of both studs established, you'll know the position of the top plate, as well.

To determine the difference in length between the shortest and longest studs in a rake wall, you need to know both the length of the wall and the pitch of the roof. For example, in a house that's 33 ft. wide, a rake wall running to the center of the roof is 16 ft. 6 in. long. With a 6-in-12 roof pitch, multiply 6 by 16 ft. 6 in. (6 in. of rise for every foot of run and 16 ft. 6 in. of run) for a result of 99 in. Add 99 in. to the length of your shortest stud, and you've got the length of your longest stud—191¼ in.

Now go back to the subfloor to lay out the top plate (drawing next page). First, go to the end of the chalkline marking the bottom plate, where the plan indicates the low point of the rake wall. Usually this will be at an exterior wall, but check the plans for the exact location of the shortest stud. Measure up 92¼ in. on the subfloor and mark that point. Next,

From *Fine Homebuilding* magazine (February 1992) 72:67-69

come over 16 ft. 6 in. along the same chalkline to the house's center. Measure up 191¼ in. from there for the long stud and mark that point. Make sure your measurements are perpendicular to the chalkline. Connect the two points with a chalkline, and you've established the location of your top plate. Intermediate studs can now be cut to length without any further calculations.

No math, no sweat—Not every good carpenter tackles problems this way, however, and calculators still haven't become commonplace in most tool belts. Another method of laying out rake walls, developed by framers, dispenses with calculation altogether. The trick is to work on a 12-ft. grid and to figure the pitch in feet rather than inches (drawing facing page).

Let's look at the same problem again: a 6-in-12 pitch and a 16-ft. 6-in. wall. Measure up 92¼ in. from the bottom-plate chalkline at the low end of the rake wall. Mark that point; the height of the short stud hasn't changed. Next, come over 12 ft. along the bottom-plate chalkline and again measure up 92¼ in. perpendicular to the chalkline. Mark this point. So far, all you've done is lay out a rectangle that is 92¼ in. on the short sides and 12 ft. on the long sides.

From the last point, at the 92¼-in. mark, measure straight up in feet whatever the roof pitch is in inches. In this example, because you're working with a 6-in-12 roof pitch, measure up 6 ft. (the rise for a 12-ft. run) and mark that point. Snap a line between this point and the top of your short stud, and you've got your roof pitch. Because your wall is longer than 12 ft., it's necessary to extend this top-plate chalkline several feet. Complete the layout by snapping a chalkline that will represent the outside edge of the longest stud at 16 ft. 6 in. You now have a full-size layout of the rake wall. This process works regardless of the pitch of the roof or the length of the wall and usually can be completed in just a few minutes.

You may run into situations where there isn't enough floor space to lay out the rake wall using this method. When that occurs, simply cut the wall length in half and lay out the pitch as a 3-in-6 instead of 6-in-12. This way, you only need 6 ft. of floor space.

Remember that the lines snapped on the floor show the length of the shortest and longest studs (at their outside edges), the

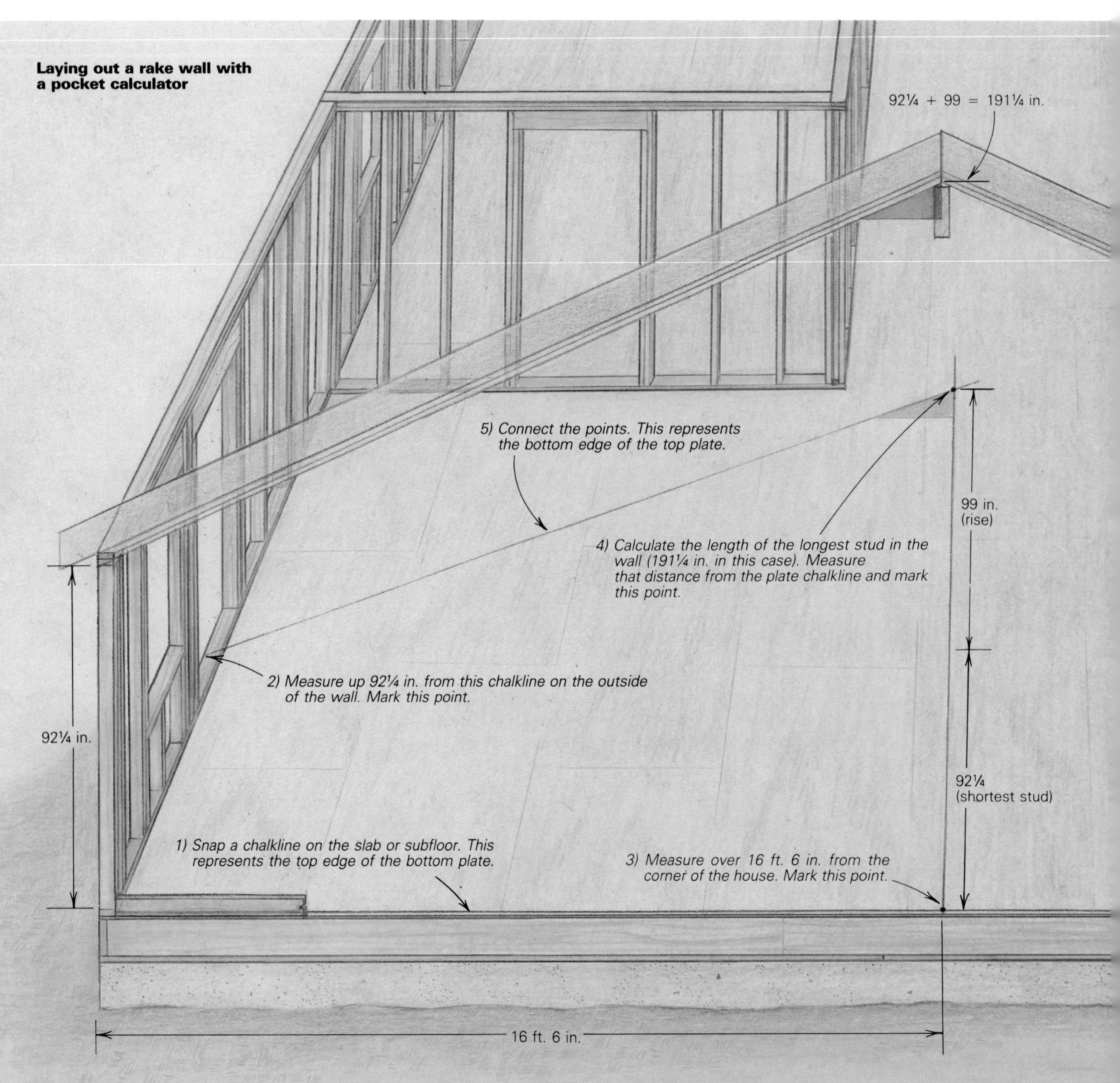

Drawings: Bob Goodfellow

roof pitch and the length of the wall. The bottom plate goes below the bottom line, and the top plate goes above the top line as the wall is framed.

Framing a rake wall—Once you've got the perimeter of your rake wall laid out, mark *two* bottom plates (you'll see why in a moment) with stud, window and door locations. Place a stud at every layout mark, letting them extend a little beyond the top-plate chalkline. Nail these studs to one bottom plate, including any trimmers and king studs. Also nail in any headers at this stage. Cripples on top of the headers need to run past the top-plate chalkline, just as the studs do. Next, position the bottom plate so that its top edge is on, but below, the chalkline. Tack it in place with a few 8d nails to make sure it stays straight, and use the extra bottom plate as a layout guide to align the loose top ends of the studs. Then pull a chalkline on the studs directly over the roof-pitch line and snap it to mark the studs for length (photo, p. 49). Before cutting the studs to length, bring in the piece of lumber that will be your new top plate, place it on edge directly above the chalkline, mark it for length and indicate on it where the studs will be nailed once they are cut.

Now it's time to cut the studs to length. If the saw's shoe tilts in the right direction to make the cut, set it at the proper degree for the roof pitch (26½° for a 6-in-12 pitch). If the angles marked on the studs run opposite the direction in which your saw tilts, first cut them square and then make the second cut at the correct angle. The next step is to nail on the top and double plates. Lap the double plate over 3½ in. at the low end to tie the two walls together at the corner.

How much precision is necessary?—It's been my experience that carpenters often spend too much time on rake walls, trying to build them to extremely fine tolerances. It usually doesn't matter if these walls get built a little high or a little low. With a site-built roof, I actually like to run the rake wall at least 1 in. high so that a good tie can be made between it and the rafter sitting atop it. □

Larry Haun lives in Los Angeles, Calif., and is a member of Local 409; he was a longtime teacher in the apprenticeship program.

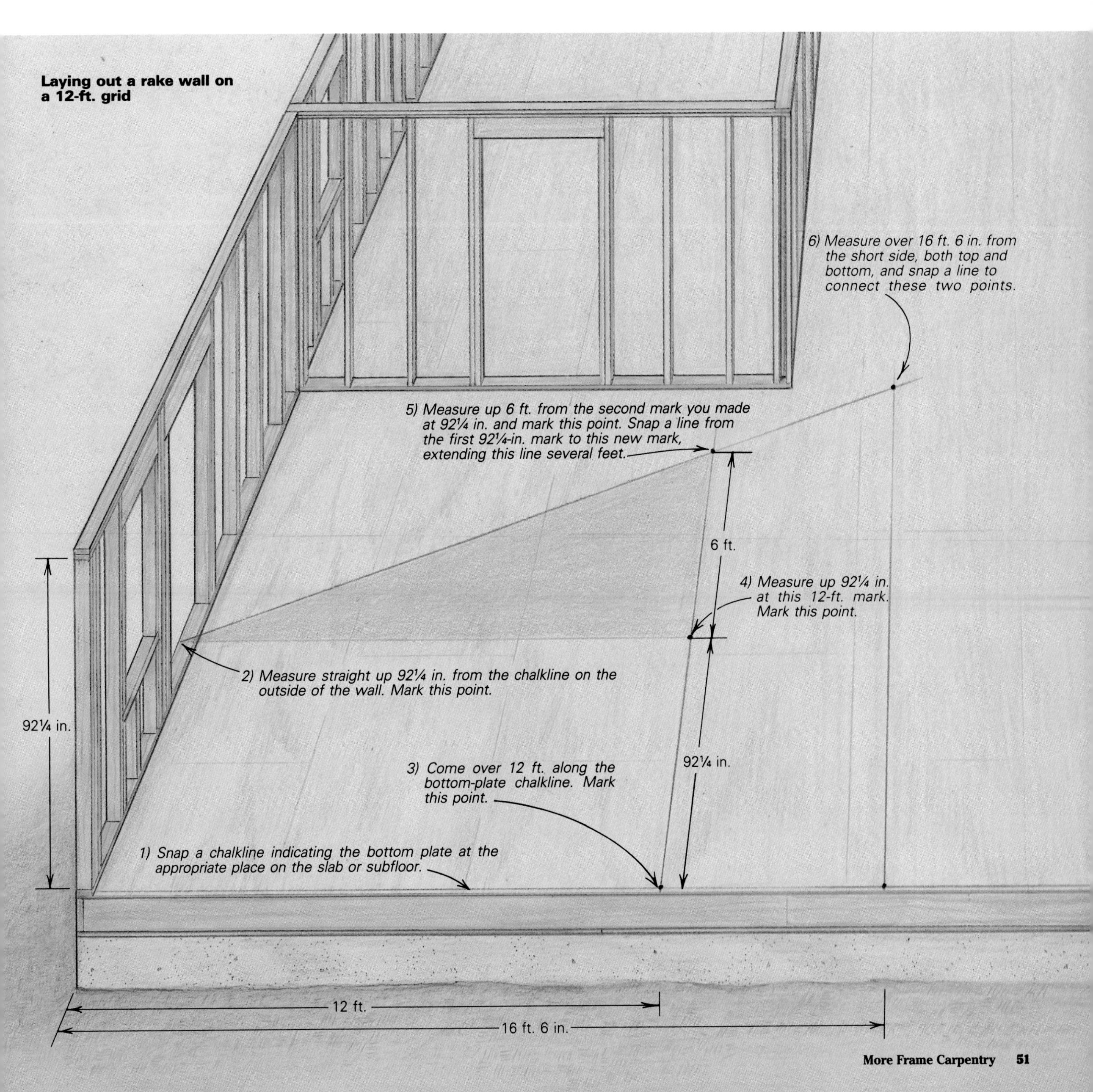

Balloon-Framing a Rake Wall

How one builder stiffens walls by running studs from floor to roofline

by Sean Sheehan

Here in Montana, the wind is something you can count on. The broad mountain valleys that grace this state are, among other things, nature's wind tunnels (the high plains are called windswept for good reason). A builder must contend with the wind during all phases of a project, and any building should be designed with the wind in mind.

One technique that our crew uses to increase the wind resistance of a rake wall (a wall whose top plate follows the incline of the roof) is balloon-framing. Whenever we can, we extend the rake-wall studs all the way from the floor to the roofline rather than frame a conventional wall and stand a truss on it or fill in above the wall with gable-end studs.

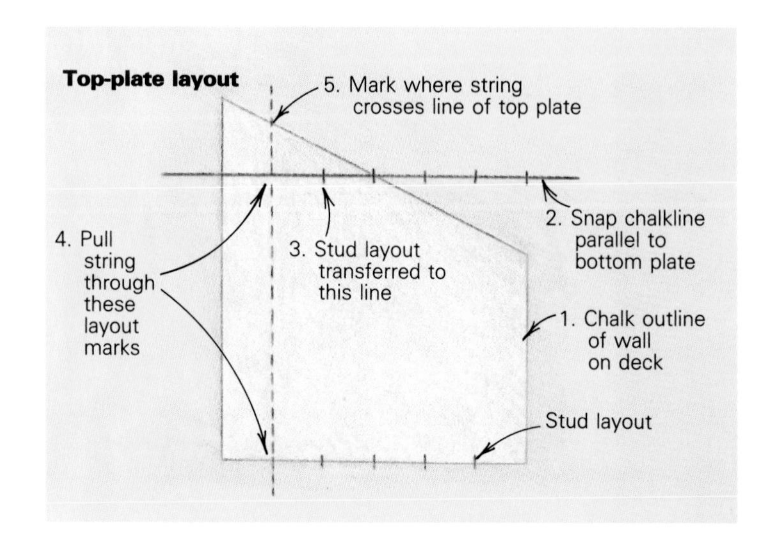

The concept

Think of the rake wall as a triangle on top of a rectangle.

Balloon- vs. platform-framing—In balloon-framing, studs run continuously from foundation to roof. The second floor, if there is one, hangs from the studs. In platform-framing, which evolved from balloon-framing as a safer and more efficient form of construction, the second floor is built on top of the first-floor walls. Then the second-floor walls are framed on top of the platform (hence the name). With this system the top plates of the first-floor walls serve as firestops; in balloon-framing, firestops have to be added. Platform-framing also requires shorter studs, which are easier to handle, and provides a safe platform (the second floor) on which to work, rather than requiring the carpenter to build walls 16 ft. in the air.

Nonetheless, there are times when balloon-framing makes sense. I consider it essential when building a tall, window-filled rake wall in a home with high cathedral ceilings. Even when sheathed with plywood, the platform-framed version of this wall can literally billow in the wind. The top plates that divide a platform-framed wall from the rake-wall studs above it create a break line. When the wind pushes against such an arrangement and there is no interior structure (such as a second floor or a partition wall) to resist it, the wall flexes at the break line.

The structural integrity of balloon-framing can be undermined by careless placement of windows and doors. We try to ensure that studs in the center third of a rake wall are left intact. If this is impossible, we double up king studs, or sometimes triple-stud the center of a wall if there are windows on both sides of center. The object is a stiffer wall, so we leave enough of the studs in one piece to achieve this goal.

We use two basic methods to balloon-frame a rake wall. If we have the space, we build it on the first-floor deck. We usually divide a peaked wall into two wedge-shaped walls and nail them together after the walls are up. This provides more work space on the deck and puts a double stud in the center that runs to the peak. If we don't have room on the deck, then we build the wall in place, or "in the air," as we call it.

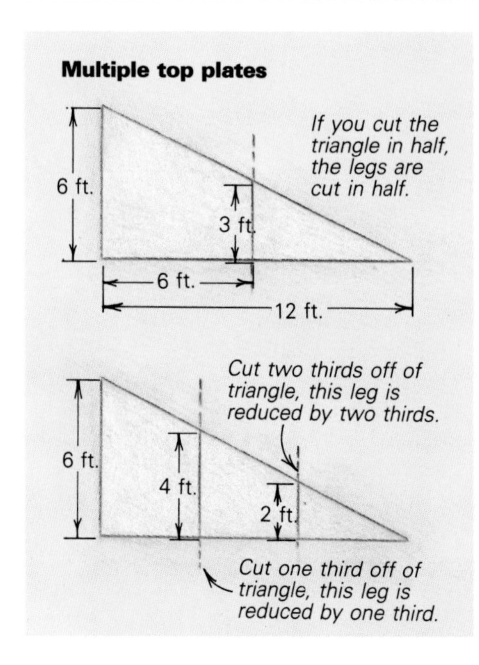

First, the math—Usually, the eave height and the length of the wall are known, and I have to determine the peak height, the length of the top plate and the length of each stud. The peak height is determined by the pitch of the roof we're building. Let's assume a 6-in-12 pitch, a wall length of 12 ft. and an eave height of 8 ft. Over 12 ft., a 6-in-12 pitch will rise 6 ft. Add 6 ft. to the height of the eave for a peak height of 14 ft. To find the length of the top plate, it's helpful to think of the wall as a right triangle on top of a rectangle (center left drawing). The rectangle is 12 ft. by 8 ft. (the length of the wall by the height at the eave), and the right triangle is 6 ft. high and 12 ft. long with an unknown hypotenuse (the top plate). I use the Pythagorean theorem ($a^2 + b^2 = c^2$) to find the length of the top plate.

In this case, $a = 6$ (the length of one leg) and $b = 12$ (the length of the other leg). Plugging these numbers into the formula, you'll find that c^2 must be equal to 180. Now all you need is to find the square root of 180, which is 13.416 ft., or 13 ft. 5 in.

On the deck—Once I have the length of the top and bottom plates and the height of the wall at both ends, I chalkline a full-scale pattern of the wall on the deck, being careful to keep the pattern square.

To mark the stud locations on the top plate, another carpenter and I snap a line parallel to

Drawings: Michael Mandarano

the stud layout of both the bottom plate and this line. We mark the points where the string crosses the line of the top plate.

Once the layout is accurately transferred to the top plate, it's a simple matter to measure stud lengths. We refer to the measurement as being to the "long side" or the "short side" to avoid confusion, and if we're building more than one wall from the pattern, we write the measurements on the deck below each stud.

In the air—When there is no deck on which to lay out the wall, we take a different approach. If the wall is short enough that the top plate can be cut from a single piece of lumber, we simply cut the plates and the end studs, nail these together and erect this frame. Once the frame is up and braced plumb, we lay out the studs on the bottom plate. If the wind isn't blowing, the layout can be transferred to the top plate with a plumb bob (photo right).

If the wind *is* blowing, we nail a 2x4 horizontally across the outside of the frame, level with the top of the shortest stud. We then transfer the stud layout onto this. Next we stretch a string from the bottom plate layout, through the layout on the 2x4, to the top plate and mark where the string crosses the top plate. If we don't trust the straightness of the top plate (and we never do), we pull a string along its top and use a temporary stud to correct the bow.

If the wall is long enough to require a two- or three-piece top plate, the wall can be equally divided. The lengths of the studs that will stand beneath the breaks in the top plate can be determined easily with a little math.

Let's return to the hypothetical wall: the length is 12 ft., the shortest stud is 8 ft., the longest stud is 14 ft., and the top plate is 13 ft. 5 in. If we were to break the top plate into two equal pieces 6 ft. 8½ in. long, the length of the stud that would stand under this break in the plate would be equal to half the difference between the length of the shortest and the longest stud, plus the length of the shortest stud, or 11 ft. Again, it helps to use the triangle/rectangle analogy. Simply put, if you cut the triangle in half, the legs will also be cut in half (bottom drawing, facing page). This works with any division.

When this method is used, the wall usually ends up with an extra stud in the center because the layout almost never coincides with the exact center of the wall. Sometimes the center stud interferes with the installation of another stud, but still doesn't fall on the layout. In this case, we simply add another stud onto the side of the center stud closest to the layout. Keep in mind that if you want the break to fall in the center of the stud, the measurement you arrive at mathematically will be to the center of the angled cut at the top of the stud. This is the only time we deal with a measurement that is neither the "long side" nor the "short side."

Opposite sides of a peaked wall should be identical, and when building "in the air," measurements can be transferred from the top plate on one side to the top plate of the other. If things start coming out "a little off," find out why. Geometry is an exact science, and if the studs that fit on one side are suddenly ¼ in. too long on the other, resist the temptation to just squeeze them in, or push them over and figure it's close enough. Chances are good that ¼ in. isn't nearly close enough.

It's very important to maintain close tolerances when balloon-framing. Particularly on steep pitches, an error in stud length of ⅛ in. can cause a considerable bow in the top plate. This is also true with regard to placing studs on center, and to a lesser extent, it holds true with top plate length. An error of ½ in. can really throw things out of whack. Double-check your math to be sure the figures are right. Once the wall is up, we usually tack the bottom plate in a few places, brace it and immediately plumb it. When the ends are plumb, we run a quick check on each stud with a 5-ft. level.

Finally, when the wall is permanently nailed and braced, we snap a line the length of the wall at 8 ft. from the floor to serve as an installation reference for the fire blocking. In the event of fire, this will prevent a wall cavity from behaving like a chimney and increasing both the rate of spread and intensity of the conflagration. We position the blocks in an alternate pattern: one above the line, the next below, so we can nail through the studs and into the blocks. There is a danger of bowing the studs with overdimension blocks, so here again, we maintain a high standard of accuracy. □

When there isn't room on the deck, Sheehan's crew builds rake walls "in the air," standing up the basic frame, then filling in the studs. After laying out the bottom plate, they use a plumb bob to transfer the layout to the top plate, provided the wind isn't blowing too hard.

Sean Sheehan is a builder in Basin, Montana. All photos by the author.

From *Fine Homebuilding* magazine (August 1990) 62:62-63

Strengthening Plate-to-Rafter Connections

It may be time to abandon the time-honored toenail

Failure of a toenailed connection.

by Stanley H. Niu

Overloaded rafter-tie connection.

Overloaded lag-screw connection.

Late September 1989: Hurricane Hugo clobbers South Carolina. April 26, 1991: Tornadoes knife through Butter County, Kansas. In these and many other instances of extreme weather, wood-frame houses are among the most heavily damaged structures. The mode of failure is predictable: The roof blows off, leaving bare walls to weather the storm.

Many people consider the damage caused by hurricanes and tornadoes to be an act of fate and assume that nothing can be done to prevent the destruction. Perhaps this is true, but I'm convinced that the damage can certainly be reduced, and with minimal expense. If the roof

From *Fine Homebuilding* magazine (April 1992) 74:36-39

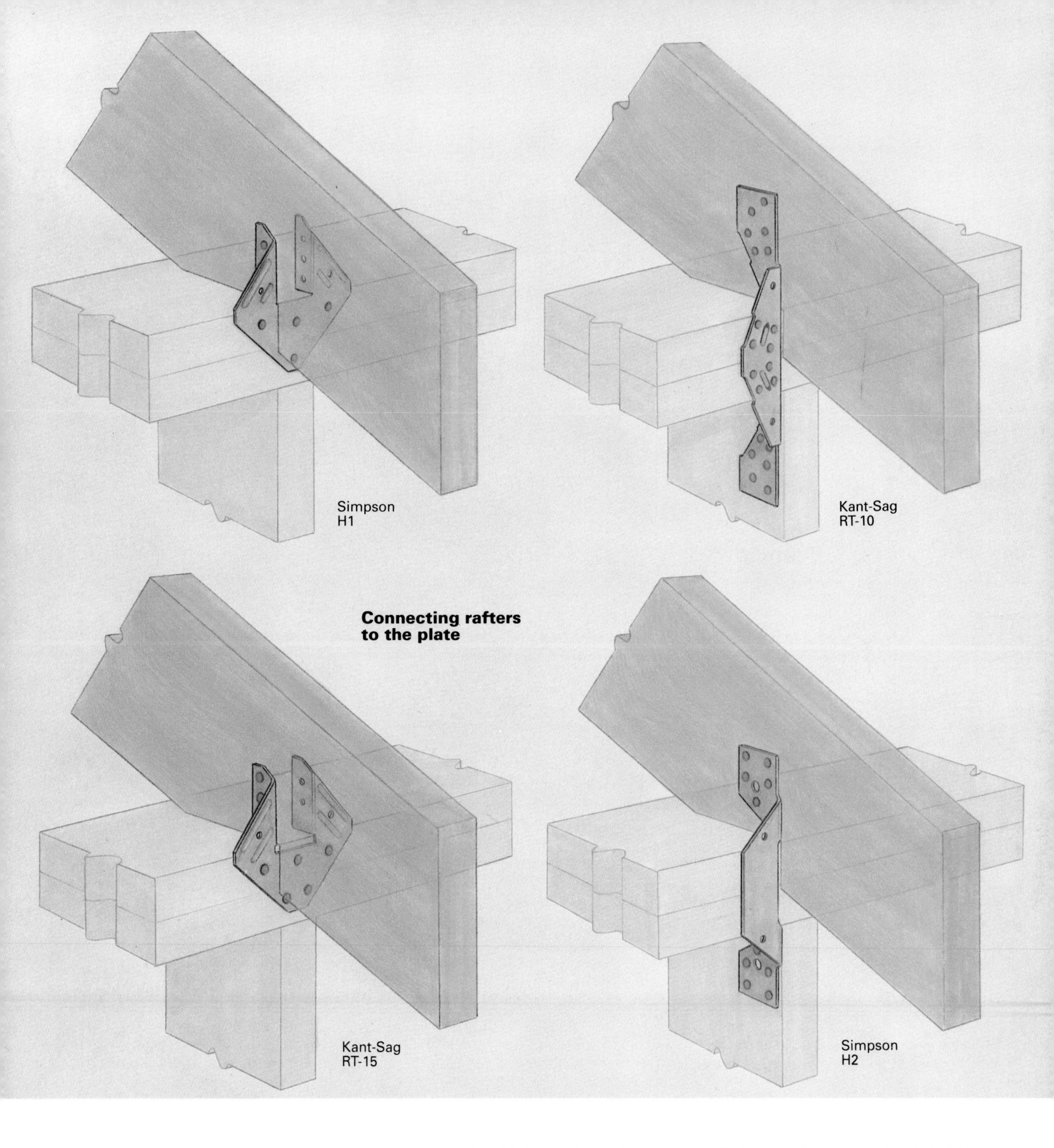

stays in place, the rest of the house stands a better chance of resisting the storm. The key is to improve the strength of the connection between the top plates and the rafters. Research I recently undertook with colleagues Laurence Canfield and Henry Liu shows just how much this connection can be improved.

Making a better connection—Although wind pressures on buildings have been studied extensively, only a few studies have examined the strength of the rafter/plate connection. What is known, however, is that a connection made with metal rafter ties is considerably stronger than one made by toenailing. Unfortunately, not all manufacturers publish information regarding the maximum recommended uplift loads their ties can resist. Without those figures, it's tough to pick the appropriate tie. So in our laboratory, we tested a selection of ties in various shapes from several manufacturers (chart, p. 57) to establish the ultimate strength of each tie. We also investigated the uplift resistance of toenailed connections, as well as two different sizes of lag-screw connections (the lags were run through the rafter and 3 in. into the plates; a washer was included).

First, I'll give a couple of notes about our testing procedures. Nails used for the three different toenailed connections we tested included 8d common nails, 16d box nails and ring-shanked, 16d common pole barn nails. The 16d nails often split the rafter during nailing, so we predrilled the rafters with a $\frac{5}{32}$-in.-dia. hole. It is unlikely, however, that carpenters would drill pilot holes in the field. Each toenailed connection used three nails: two on one side and one centered on the other side. The lumber we used for all tests was construction-grade stock obtained from a job site, and we inspected it to ensure that no flaws or cracks would bias the test results. After the appropriate rafter/plate connection was made, samples of the assembly were placed in a

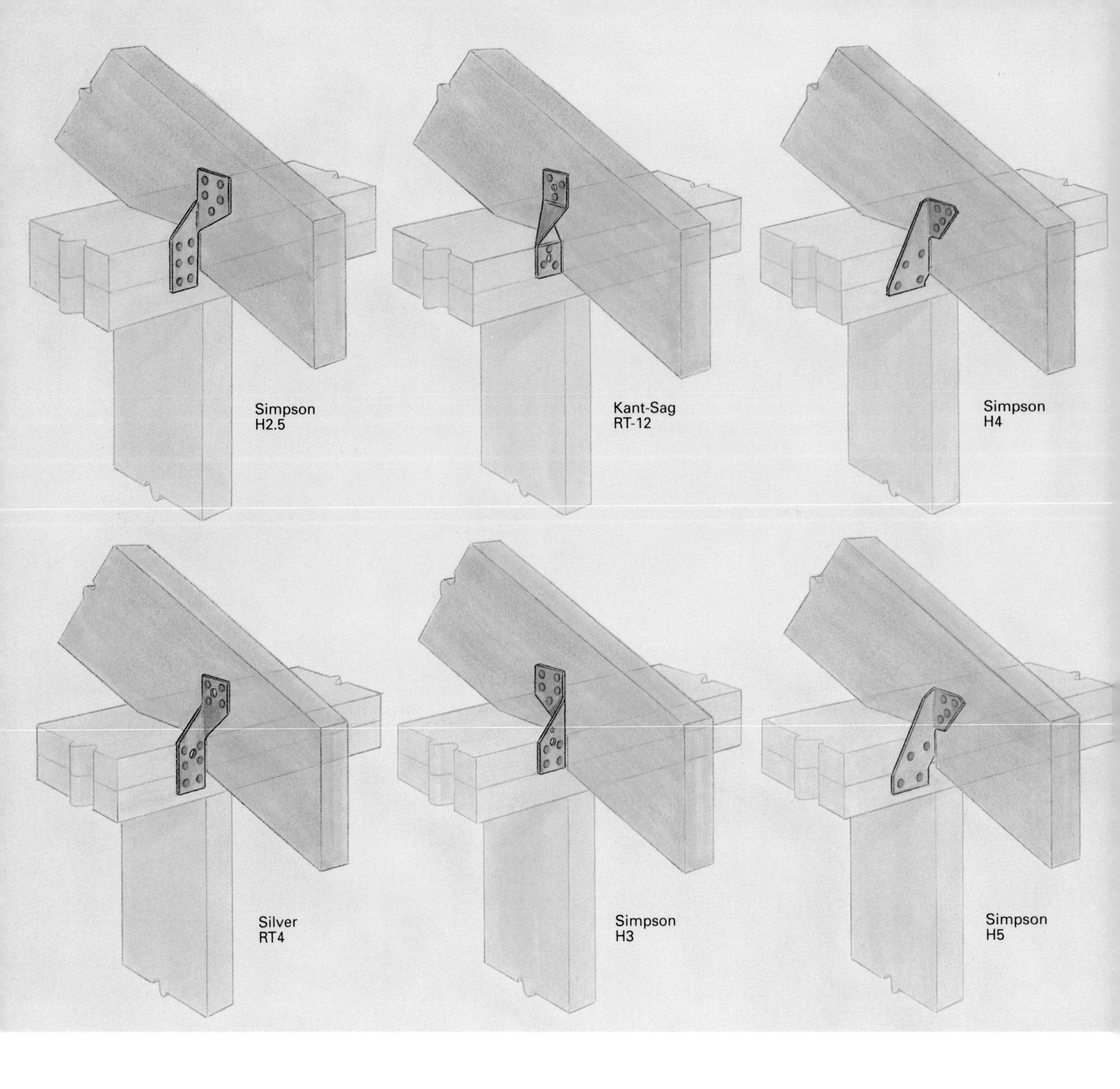

hydraulic test apparatus that pulled the rafter away from the plate. We tested at least 15 connections, pulling until the connection failed.

The results of these tests are shown in the chart on the facing page. Ties fell into three groups ranked according to their average load capacity: below 650 lb.; 900 lb. to 1,300 lb.; and above 2,700 lb. (the last group represents the high-performance end of the spectrum, with a load capacity that is double or triple most of the midrange connections). As it turns out, the weakest sample tested was the 8d toenail connection, with an average load capacity of only 208 lb. In contrast, the lowest-capacity rafter tie tested had an average load capacity of 497 lb. When the toenailed connection failed, the nails pulled out of the top plate (top photo, p. 54). In some cases, when the connection failed, the bottom of the rafter split first. When a metal tie fails, it usually tears in half, but the nails stay put (bottom left photo, p. 54). The lag-screw connections failed when the lag pulled free of the top plate (bottom right photo, p. 54). Unfortunately, toenailed rafters are probably the most common rafter/plate connection found in wood-frame houses. In fact, this connection is in compliance with the Uniform Building Code (UBC) and the Building Officials and Code Administrators (BOCA).

Applying the research—Putting our results to the test on a hypothetical house shows how important it can be to use the right connection. Consider a house located near Kansas City, Missouri, with a 30-ft. by 60-ft. floor plan and a hip roof. The rafters are located 16 in. o. c., which calls for a total of 86 rafter connections, and the roof has a 3-in-12 pitch with no overhang. The house is located on open terrain surrounded by scattered obstructions having heights of 30 ft. or less. A map of wind speeds shows that the Kansas City area has a basic wind speed of 75 mph (basic wind speed, an engineering term, is the fastest wind speed measured at 33 ft. above the ground with a 2% annual probability of occurrence).

For an 1,800-sq.-ft. roof, the total wind lift on our hypothetical house equals 31,824 lb. Dividing this by 86 connections yields a 370-lb. uplift load per connection. Based on our test results, any of the ties tested would be adequate for this region. However, a connection made with three 8d common nails has a load capacity of only 208 lb. In some weather conditions, this connection would be inadequate.

Now consider the same house located on oceanfront property in South Carolina. The basic wind speed there is 100 mph, so the load per connector would be 957 lb. on the same roof. Metal connectors from the middle group (load capacity from 900 lb. to 1,300 lb.) would be adequate, though some barely so. However, wind speeds

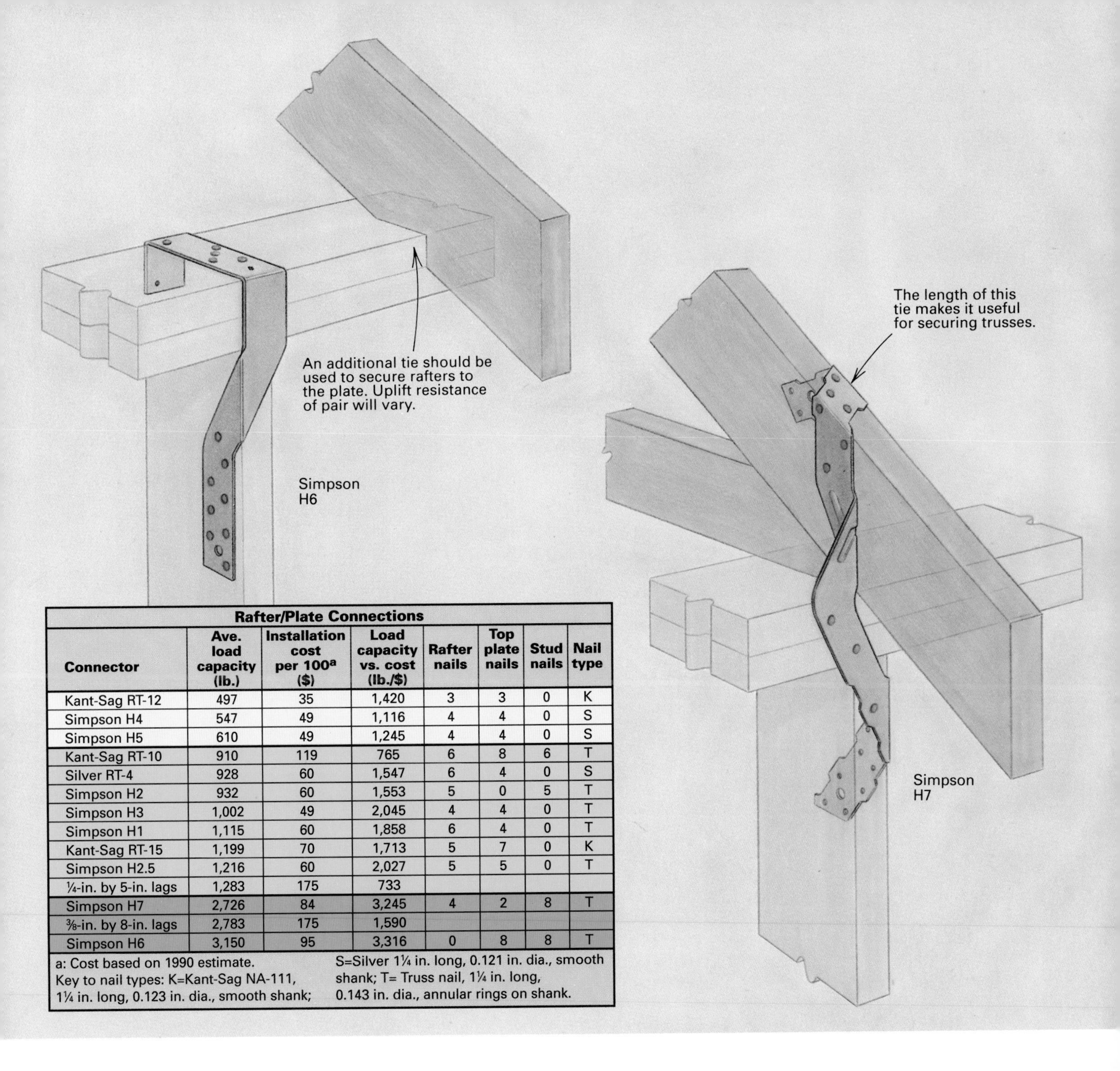

Rafter/Plate Connections

Connector	Ave. load capacity (lb.)	Installation cost per 100[a] ($)	Load capacity vs. cost (lb./$)	Rafter nails	Top plate nails	Stud nails	Nail type
Kant-Sag RT-12	497	35	1,420	3	3	0	K
Simpson H4	547	49	1,116	4	4	0	S
Simpson H5	610	49	1,245	4	4	0	S
Kant-Sag RT-10	910	119	765	6	8	6	T
Silver RT-4	928	60	1,547	6	4	0	S
Simpson H2	932	60	1,553	5	0	5	T
Simpson H3	1,002	49	2,045	4	4	0	T
Simpson H1	1,115	60	1,858	6	4	0	T
Kant-Sag RT-15	1,199	70	1,713	5	7	0	K
Simpson H2.5	1,216	60	2,027	5	5	0	T
¼-in. by 5-in. lags	1,283	175	733				
Simpson H7	2,726	84	3,245	4	2	8	T
⅜-in. by 8-in. lags	2,783	175	1,590				
Simpson H6	3,150	95	3,316	0	8	8	T

a: Cost based on 1990 estimate.
Key to nail types: K=Kant-Sag NA-111, 1¼ in. long, 0.123 in. dia., smooth shank; S=Silver 1¼ in. long, 0.121 in. dia., smooth shank; T= Truss nail, 1¼ in. long, 0.143 in. dia., annular rings on shank.

of 125 mph were reported during Hurricane Hugo, and lifting loads during the storm would have been 1,496 lb. per connector. Only the top two connections would have been adequate: the H7 tie and the ⅜-in. by 8-in. lag screw. Of course, the rest of the structure would require sufficient strength to prevent it from being blown off the foundation. But either of these connections would have improved the chances of keeping the roof in place.

The cost of safety—Our research was done in a laboratory, so it was easy to see which connection would be the best. But on the job site, "best" often competes with "cost-effective" for the right to determine what gets built. That's why we calculated the installed cost of each connection. In determining the costs, we assumed that a carpenter would take 10 seconds to install each nail and would earn an average wage of $21 per hour. The average house would probably require from 80 to 120 connectors. As you can see from the chart above, the additional cost incurred by using rafter ties is negligible compared to the total cost of the house.

Manufacturer's guidelines suggest that ties be installed with at least four nails each to prevent the tie from rotating. However, more nails ensure a better connection.

Improving the improvements—Though the rafter ties performed well as a group, we identified some modifications that could improve their uplift strength. The H4 and H5 rafter ties could be made from 18-ga. sheet metal instead of the 20-ga. sheet metal currently used, and the nail holes could be slightly larger to accommodate truss nails. The H2 rafter tie has a hole on its face between the rafter and the top plate. In our tests, the tie failed by tearing in half, with the tear starting on the inside edge of the tie and progressing to this hole. Elimination of the hole might improve the strength of the tie. The RT-10 rafter tie, which is similar to the H2, also failed by tearing in half between the rafter and the top plate. This tie would be improved if it were wider (more like the proportions of the H2). Generally, the 18-ga. sheet metal used for most of the ties seems a good compromise between strength and low manufacturing cost. □

Stanley H. Niu is an associate professor in the Department of Civil Engineering at the University of Missouri. Professor Henry Liu supervised the research. Laurence Canfield, plant engineer at the Wire Rope Corp. of America (St. Joseph, Mo.), conducted the experiments. Photos by the author. For further information on the methodology of these tests, see the Forest Products Journal, *July/August 1991 (2801 Marshall Ct., Madison, Wisc. 53705).*

Riding Out the Big One

A structural engineer explains what happens to a wood-frame building in an earthquake and how it can resist collapse

by Ralph Gareth Gray

The area affected by the Loma Prieta earthquake, which shook up much of the San Francisco Bay Area in October of 1989, included 1,544 California public schools. Only five of them suffered severe damage. No lives were lost, and there were no injuries in these buildings. Most of them were wood-frame construction very similar to that of a typical custom house, and they were either built from scratch or upgraded following the five sacred principles of earthquake-resistive construction.

There's nothing mysterious about these principles, and I'll discuss them all in this article. To understand them, you must first visualize clearly how a structure carries the loads imposed on it by an earthquake. To put the principles to use in the real world of construction, you might have to make a few simple but critical modifications to standard building practice. Also, you must pay close attention to details, and to get them done right, you'll probably have to be steadfastly persistent. This is a big topic, so in this article I'm going to talk mostly about new construction. In a subsequent piece we'll look at retrofitting existing houses to withstand earthquakes.

Some of what follows is subject to continuing debate among structural engineers, and I'll probably rub some of my colleagues the wrong way. But shall I tell you what I really think, or dose you with bland, safe consensus? I choose the former.

Sudden shock—In an earthquake, the ground moves violently and chaotically, up and down, side to side, twisting and rocking, with all these motions changing very quickly. Anything on the ground, such as a house, will tend to slide and overturn. Various parts will rattle around on their own and maybe come adrift. Things stacked one on another, like bricks on deteriorating mortar, will tend to slide and overturn independently, and framing members, like beams, may come right off their posts, or the posts off their footings. The connecting links—particularly tension and shear-carrying components—are put to the supreme test in an earthquake. If they are incorrectly designed or constructed, they will cause serious trouble. Thus the first principle: *Tie it together.*

Overlaps and straps—The platform frame holds together better in earthquakes than the balloon frame, in part because a platform frame is tied together by the overlap of the double top plates on the walls at every level. But they must be spliced correctly, with minimum 4-ft. overlaps (top drawing, facing page). While the code requires at least two 16d common nails on each end of the splice, I think four 16d nails is a better minimum. If more are needed, the drawings should show them. As you add more nails, follow a nailing pattern like the one shown in the drawing to avoid splitting the wood.

Post-to-beam connections are also important, and any measure of reinforcement is better than the following common condition: a couple of toenails through the bottom edge of the beam into the post's end grain. Instead, use steel connectors, such as commercially available post caps. You can make an effective site-built post-to-beam connector using a 2x bolster and a couple of 2x yokes on either side of the post to cradle the beam (bottom drawing, facing page).

Steel column bases are the best way to tie posts to their piers (drawing below, facing page). But even minimal connections, such as short steel Ls secured with drilled-in anchor bolts, or even plumber's tape affixed with concrete nails are better than a couple of toenails.

Parts other than framing members should also be tied together. Use metal straps to secure mechanical equipment, such as the water heater, to the walls. Concrete or clay roofing tiles (and roofing slates) should all be anchored to the roof sheathing with corrosion-resistant fasteners. Unfortunately, there are roof-tile systems, with ICBO approvals, that do not

From *Fine Homebuilding* magazine (December 1990) 64:60-65

require mechanical fasteners above the first few courses. I shudder to think what might happen to people who are running out the door during a serious shake when the tiles depart the roof.

Stucco can be a fine finish, and it's excellent fire protection, but it has to be tied to the wall. This means that the mesh must stand off the sheathing far enough to key and support the scratch coat, and the nails mustn't pull out of the sheathing because it's decayed or because the nails are too short, rusted or both. In San Francisco's Marina District I saw great sheets of stucco that had peeled right off the walls. If the Marina fire had spread, that stucco wouldn't have protected anything but the paving it was lying on.

Inertia—The sudden movement of the ground under the building and the rapidity of change in the ground's motion set earthquakes apart from other dynamic actions on buildings, such as wind. The sliding, overturning and so on are primarily due to inertia, a physical object's reluctance to be moved, and once moving, its reluctance to change direction or velocity.

The heavier the object, the greater its inertia. The heavier the building, the harder it is to control the forces applied by an earthquake. Thus the second principle: *Keep it light.*

If you're dead set on having a handsome veneer of brick on the sides of your house, remember that the inertial force due to its weight in sudden motion must be channeled successfully through the structure to the ground. That's because our first principle says that we have to tie it together. If you don't tie it, the force will be smaller, but only because the brick will have been thrown off the building, like great lumps of shrapnel. I have clients with expensive earthquake damage to their house that occurred because the brick veneer added to the effective weight of the building, and the structure couldn't handle the increased inertial forces.

Brick chimneys act about the same way as brick veneers, but more so. They are tall and narrow, so the whiplash tension across the mortarbed joints is significant. They are seldom tied to anything, they typically break off in an earthquake and their weight makes them deadly. I try to get people to take down the brick chimney and put in a metal prefab fireplace and double-wall metal flue. If the whole thing can't be removed for some reason, then take it down to the smokeshelf and go up from there with a metal flue. Prefab metal transition pieces are made for precisely this application. Be sure to grout it in so hot gas can't escape.

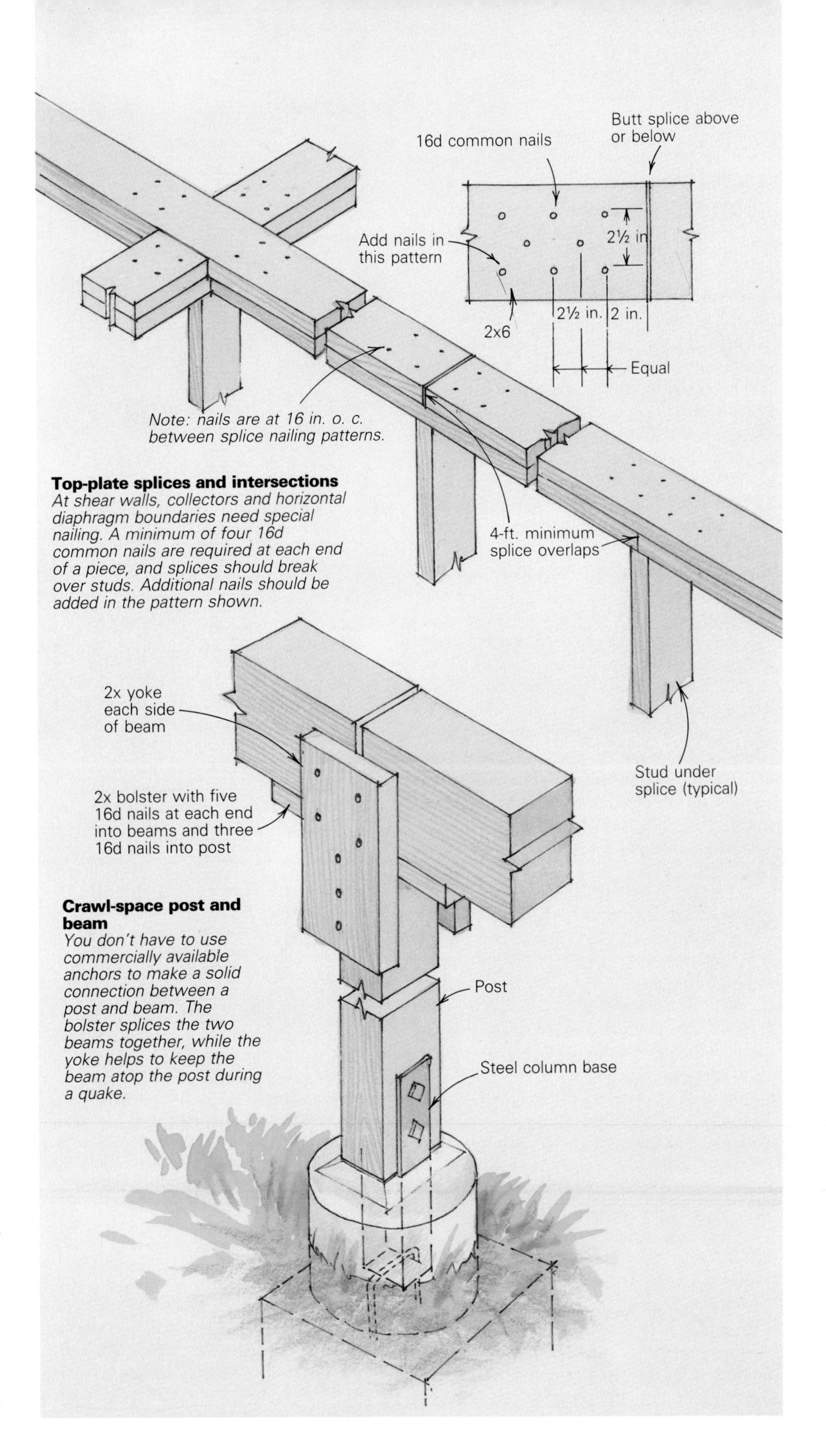

Top-plate splices and intersections
At shear walls, collectors and horizontal diaphragm boundaries need special nailing. A minimum of four 16d common nails are required at each end of a piece, and splices should break over studs. Additional nails should be added in the pattern shown.

Crawl-space post and beam
You don't have to use commercially available anchors to make a solid connection between a post and beam. The bolster splices the two beams together, while the yoke helps to keep the beam atop the post during a quake.

Tracking the loads—Another key to a building's survival in an earthquake is a strong, stable path for the transmission of inertial forces to the ground. Engineers size the fasteners, shear panels and boundary members (framing elements that take tension and compression loads) that transmit these forces based on a portion (typically around 15%) of the building's weight.

Understanding the forces at work during an earthquake can take some mental gymnastics (see load-path sidebar, next page). But when it comes to designing, building, repairing and retrofitting buildings to resist earthquakes, being able to track the inertial forces through their load paths is an essential skill. Therefore our third principle: *Know the load paths.*

If you can see a plausible path for loads to travel when you study the potential shaking of a house in each of the principal directions, you are a long way toward an effective earthquake-resisting system, and with this knowledge, you're also in a position to understand some of the more subtle but still important aspects of anti-earthquake construction.

As you can see from the drawings of our little house (see drawings in sidebar), the shapes of its various parts change during the violent shaking of an earthquake. If there's anything in the way, like a tree, the house next door, or even parts within the same structure, it gets hit with a lot of energy—sometimes repeatedly. While this battering dissipates energy, it can cause big chunks to fall off a house and even cause an otherwise sound building to collapse. This leads to our fourth principle: *If you can't tie it together, separate it.*

Obviously we can't separate elements that depend on each other for load transmission, such as a beam from its post. But what about separate portions of a house that might tangle during an earthquake? Consider, for example, a split-level house that steps down a hillside. Assume that the lower part of the house has a big roof abutting a stud wall that carries the weight of another roof (the one over the uphill portion of the house). During an earthquake, the lower roof will try to shake at a different frequency. Thus, the studs might have to be huge—maybe 3x8s at 12 in. o. c.—because in an earthquake, the roof would push and pull against them with great force. In a case like this, it could make sense to separate the lower roof and the walls from the upper wall by 2 in. or so (by about ¼ in. per foot of wall height), which means building a separate studwall to support the lower roof rather than nailing it into the studwall supporting the upper roof. Between the upper wall and the lower roof, you should provide flashings with slip joints so the elements can shake around in all directions without tearing.

A sewer, gas or water pipe passing through a wall and then turning down into the ground is locked into the wall if there isn't space around it at the wall. Leave a ½-in. gap around all sides of such lines, and fill around them with flexible caulk.

Myth of the fantastic optimum—Building structures are analogous to chains, with each element serving as an appropriately sized link to carry out its function. In a perfect world you could design and build a house using the very minimum number of fasteners, the smallest allowable foundation and the lightest structural members. Unfortunately, that's an unsafe and unrealistic approach, because the loads from a strong earthquake are just plain unpredictable. This leads to our fifth (and most important) principle: *Buildings should fail gently.*

Notice that our model building has a shear wall on each side of the floor diaphragms, for a total of four. It's possible to get by with three, provided they are laid out correctly (they should never align to a point, like spokes in a wheel). But even with three, if you lay them out correctly, there's no reserve for construction or design errors, dry rot or future remodeling by klutzes. If one of those three walls fails, for whatever reason, the whole house will fail. This is a plea for redundancy in the structure—the provision of more than just the minimum load path. Put some plywood on other walls (not shown on our drawing), all the way up to the roof, so that if one element fails, the loads flowing through the structure

(Text continued on p. 62)

Load path

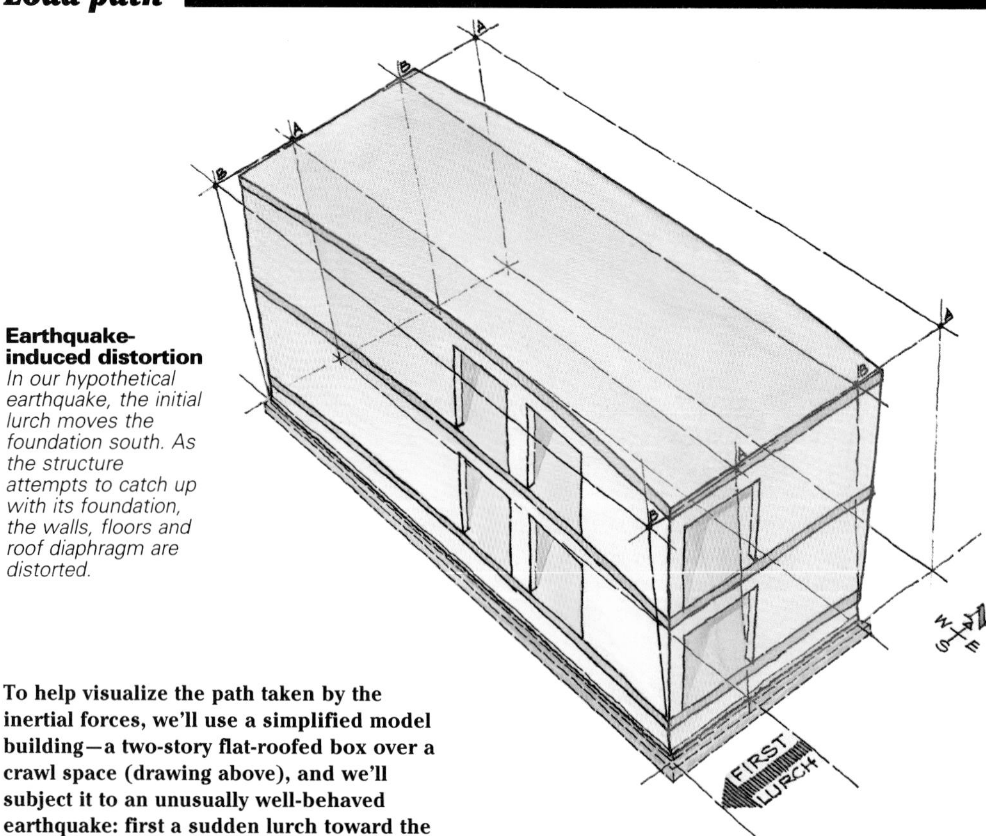

Earthquake-induced distortion
In our hypothetical earthquake, the initial lurch moves the foundation south. As the structure attempts to catch up with its foundation, the walls, floors and roof diaphragm are distorted.

To help visualize the path taken by the inertial forces, we'll use a simplified model building—a two-story flat-roofed box over a crawl space (drawing above), and we'll subject it to an unusually well-behaved earthquake: first a sudden lurch toward the south, followed by a sudden lurch back to the north. Period. We'll start at the bottom of the building and go up, concentrating on primary effects.

***The first lurch*—You can see where the house starts from by the ghost outlines defined by the A labels on the corners of the roof. During the first lurch, the building would immediately follow the ground motion exactly to the ghost outlines labeled B on the roof corners if it weren't for inertia. The building's inertia causes it to lag behind the ground, as shown by the rendering.**

A good foundation is embedded in the ground, so during the first lurch it is carried south along with the ground, taking the jack walls around the crawl space with it because their mudsills are anchor-bolted to the footings. The tops of the walls lag behind a bit, and the east and west walls are changing shape from a rectangle to a parallelogram—a kind of distortion called shear distortion—hence the name shear wall. If they were weak, the east and west walls would just mash over to the north. But they are sheathed with plywood (or diagonal sheathing) and are thus stiff and strong. Unsheathed jack walls are a common weak link in older wood-framed buildings.

The jack walls on the north and south are also sheathed, but can't resist north-south motions at all, for their stiffness and strength work only in an east-west direction. They follow along because their tops are nailed to the first floor, rotating as if on a piano hinge along the sill.

I've shown jack walls here because they're often used on the West Coast to elevate first-floor joists above grade and to level floors on hillsides. The same hinge principle described here, however, also applies to floor assemblies that bear directly on mudsills and stem walls.

The inertial force of the bottom half of the north and south jack walls goes to their footings and the force from the top half goes to the edge of the floor. So for this south lurch, the north and south walls are part of the load, not part of the resistance.

The east and west jack walls drag the first floor (and everything above it) toward the south, but reluctantly. That's because the first floor has its own inertia, plus some contributed by the north and south walls, and any attached partitions or mechanical equipment.

The middle of the floor lags behind its ends, the north and south edges changing from straight to curved. The south edge gets a little shorter and the north edge gets a little longer—a typical sign of bending distortion. Beams and girders act in bending (as well as shear), and that's what the floor is: a big, flat beam that resists lateral forces in a horizontal plane. It's usually called a horizontal or floor diaphragm.

The next layer of the cake, first-story walls plus second-story floor, acts much the same way. So does the top layer—the second-story walls plus the roof.

To summarize from the top down, the roof diaphragm carries its inertial load to the east and west second-story shear walls. The drawing labeled "Distorted horizontal diaphragm" (drawing right) shows a plan view of how the floor might look in mid-lurch. The little arrows represent the inertial loads from the floor itself, and those delivered to the floor from the north and south walls. Bigger arrows at the east and west ends represent the accumulated inertial loads delivered by the east and west shear walls from above. The largest arrows represent the resistance of the shear walls below, and ultimately, the foundation. They are equal to the sum of

Drawing this page: Ralph Gareth Gray

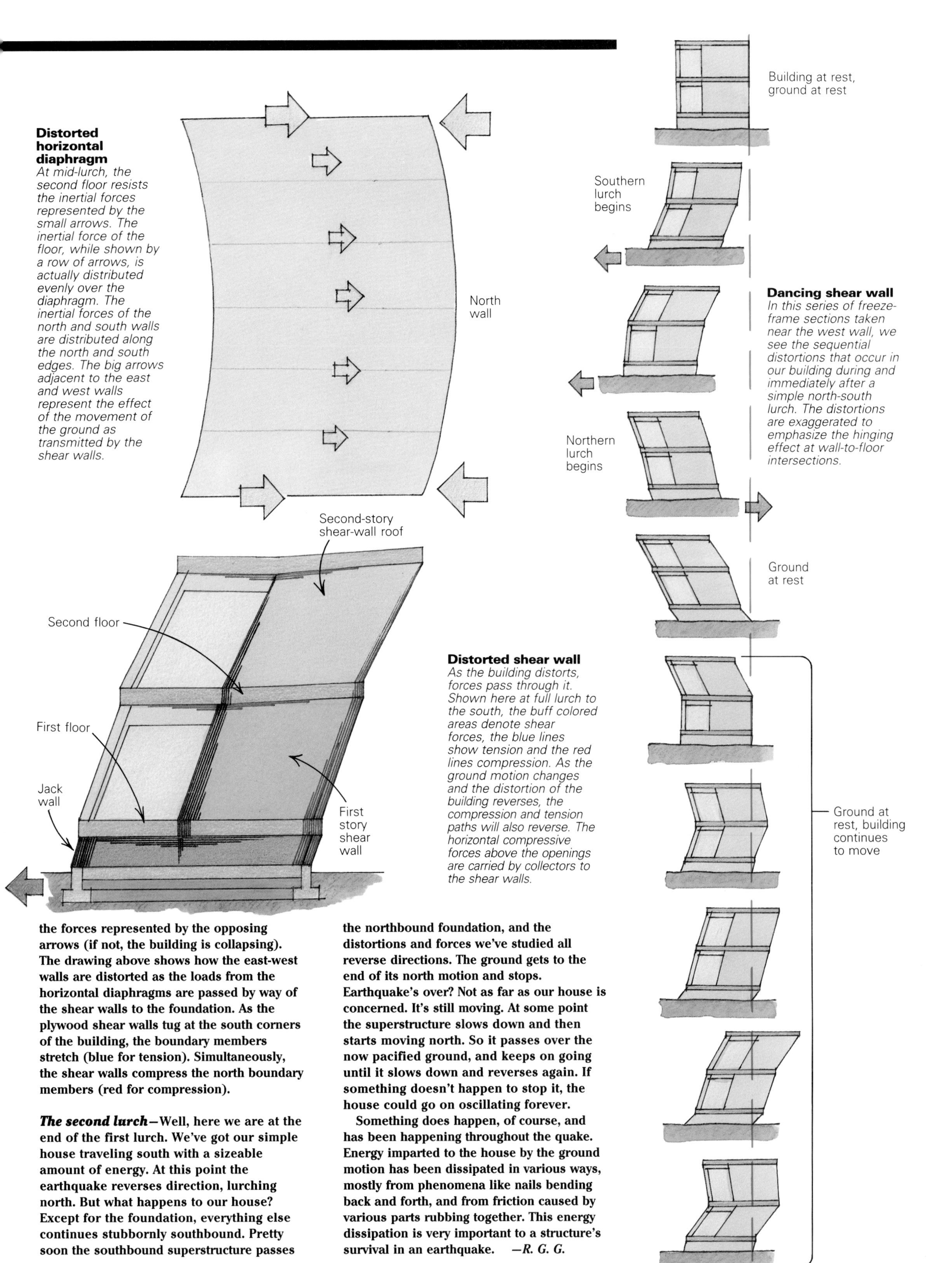

the forces represented by the opposing arrows (if not, the building is collapsing). The drawing above shows how the east-west walls are distorted as the loads from the horizontal diaphragms are passed by way of the shear walls to the foundation. As the plywood shear walls tug at the south corners of the building, the boundary members stretch (blue for tension). Simultaneously, the shear walls compress the north boundary members (red for compression).

***The second lurch*—Well, here we are at the end of the first lurch. We've got our simple house traveling south with a sizeable amount of energy. At this point the earthquake reverses direction, lurching north. But what happens to our house? Except for the foundation, everything else continues stubbornly southbound. Pretty soon the southbound superstructure passes the northbound foundation, and the distortions and forces we've studied all reverse directions. The ground gets to the end of its north motion and stops. Earthquake's over? Not as far as our house is concerned. It's still moving. At some point the superstructure slows down and then starts moving north. So it passes over the now pacified ground, and keeps on going until it slows down and reverses again. If something doesn't happen to stop it, the house could go on oscillating forever.**

Something does happen, of course, and has been happening throughout the quake. Energy imparted to the house by the ground motion has been dissipated in various ways, mostly from phenomena like nails bending back and forth, and from friction caused by various parts rubbing together. This energy dissipation is very important to a structure's survival in an earthquake. —*R. G. G.*

All other drawings: Malcolm Wells

will have an alternate path. The house might sag or show cracks but won't collapse, killing someone. Such precautions will cost a few extra dollars, but that's cheap insurance.

The horizontal diaphragm—During an earthquake, the floor and roof diaphragms undergo shear and bending. The subfloor sheathing carries the shear, like the web of an I-beam, while the boundary members (rim joists and top plates) carry the tension and compression due to bending like I-beam flanges.

To withstand and transfer the shear loads, plywood sheets have to be spliced together to prevent adjacent edges from sliding past or over each other. Plywood sheet edges should be butted and nailed to joists in one direction, and to solid blocking or rim joists in the other. Butted on the centerline of a 2x joist, you've got only ¾-in. bearing for each piece, so the nail has to be ⅜ in. from the edge. The edge-nailing called for by code can be as close as 3 in. o. c. This layout works, but there is no margin for error. Layouts must be accurate, and the nailing has to be done with care to avoid shiners and split joists or blocking.

Plywood or diagonal board sheathing is edge-nailed to the rim joists or blocking on all four sides of the diaphragm. They in turn must be connected to the top plates below, which serve as chords (like I-beam flanges), carrying tension or compression.

The edge joists and top plates compose the boundary members. Walls are often longer than the lumber available, so top plates or edge joists must be spliced for the tension and compression. If they've been severed, boundary members must be spliced (plumbers are especially good at finding and disabling the most critical diaphragm chords, usually because designers have given them no alternative). If for instance, a vent stack has been let into the side of a top plate, the load in the top plate becomes eccentric, magnifying the stress on it during an earthquake. This may snap it. Custom-made splints of ¼-in. steel angle, for example, may be required to fix this.

Large openings, like stairwells, need boundaries around them. Put blocking and strapping perpendicular to the joists across several joist spaces to compensate for the local increase in shear and bending due to the hole (drawing right). The same kind of detailing is needed at inside corners of L-shaped or more complicated diaphragms.

Shear walls and collectors—Just as the horizontal diaphragm is a big, flat beam that resists lateral forces in its horizontal plane, the shear wall is a big, flat vertical cantilever beam that resists lateral forces in its vertical plane. Again, the plywood or diagonal sheathing carries the shear, and the boundary members—stud corners or end posts—carry the bending tension and compression. Shear walls tend to be harder to engineer than floor and roof diaphragms, in part because they're smaller (one does need doors and windows). Also, loads accumulate from the top down, so the loads tend to be just plain bigger. All plywood edges, horizontal and vertical, should bear on, and be nailed to, studs, plates or blocking.

The connection between a shear wall and its foundation typically serves two functions: the transfer of shear forces delivered by the wall to the ground by way of the foundation; and the transfer of overturning forces (called uplift) to the foundation. Anchor bolts take care of the shear at the foundation level—the larger the shear force, the larger or more closely spaced the bolts. Tiedown anchors, bolts, straps or other devices resist uplift.

The shear forces from the roof boundary members are transferred to the top of the shear wall in several ways. They pass by way of nails that slant from the edge joist or blocking into the top plate, or by flat blocking between the joists nailed in turn to the top plate (the flat blocks can also be used for drywall backing). Another method is to run the wall plywood an inch or so up onto the blocking or edge joist (drawing facing page). Here, the heavy edge-nailing schedule for the wall will be used, and another line of nails will be embedded into the center of the top plate. This detail has many forms. Note that the plywood does not run all the way to the top of the joists and blocks. That's because they will shrink, while the plywood does not, causing humps in the walls and sometimes splitting the boundary members. At the joists between floors, the bottom edge of the plywood should not get too close to the top edge of the wall plywood below, because the cross-grain shrinkage of the floor or contact during the quake will strip the plywood off, split the floor members, or both. Leave about a 1-in. gap between them—enough to account for shrinkage.

At the base of each wall, shear is transferred from the plywood to the sole plate by nails, and then from the sole plate to the floor plywood. Special nailing schedules sometimes apply to this connection.

Shear walls in the lower stories resist accumulated shear, uplift and compression from the walls stacked above. Think of it all as one wall, continuous from foundation to roof, spliced at intersecting floors.

As shear forces move through walls, they have to take a path around openings for windows and doors. The forces are concentrated in the boundary members over the openings before they can be dumped into the shear wall. The framing members that handle this task are called collectors, or drag struts. Often the top plates above the headers serve this function—another reason why the top-plate splices are important. Sometimes beams are used as collectors. This is a tricky little item that is very important and often overlooked, so be aware when there are long openings in the plane of a shear wall, particularly if the roof or floor above is flush-framed. Like top plates, collectors are targets cherished by plumbers.

Fasteners—No structural element is better than its connections, and no connection is better than the fasteners. The common nail is a magnificent device for resisting earthquakes. Tests I helped conduct at the University of California's structural research facility showed that properly nailed plywood shear walls have an amazing capacity to resist earthquakes and dissipate energy owing to the ability of the nails to flex back and forth repeatedly without breaking.

But, the plywood walls secured by *overdriven* nails (nails that penetrated the plywood beyond the first veneer) failed suddenly in our tests, and at loads far below those carried by

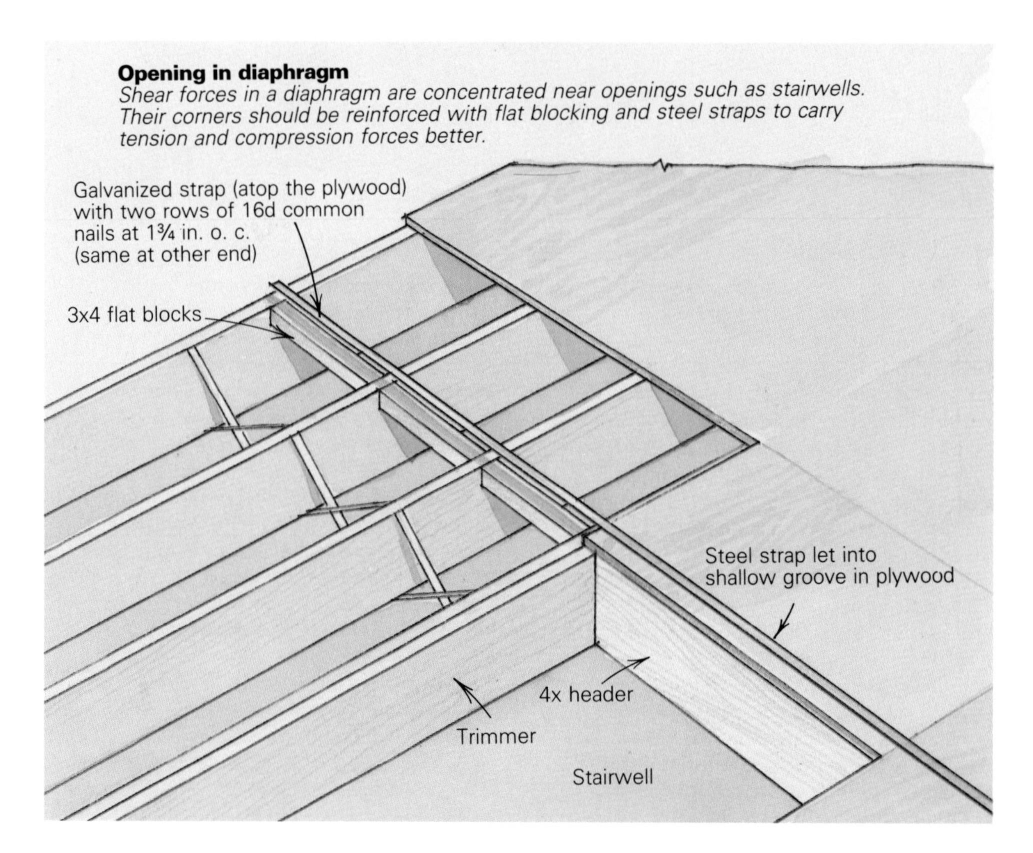
Opening in diaphragm
Shear forces in a diaphragm are concentrated near openings such as stairwells. Their corners should be reinforced with flat blocking and steel straps to carry tension and compression forces better.

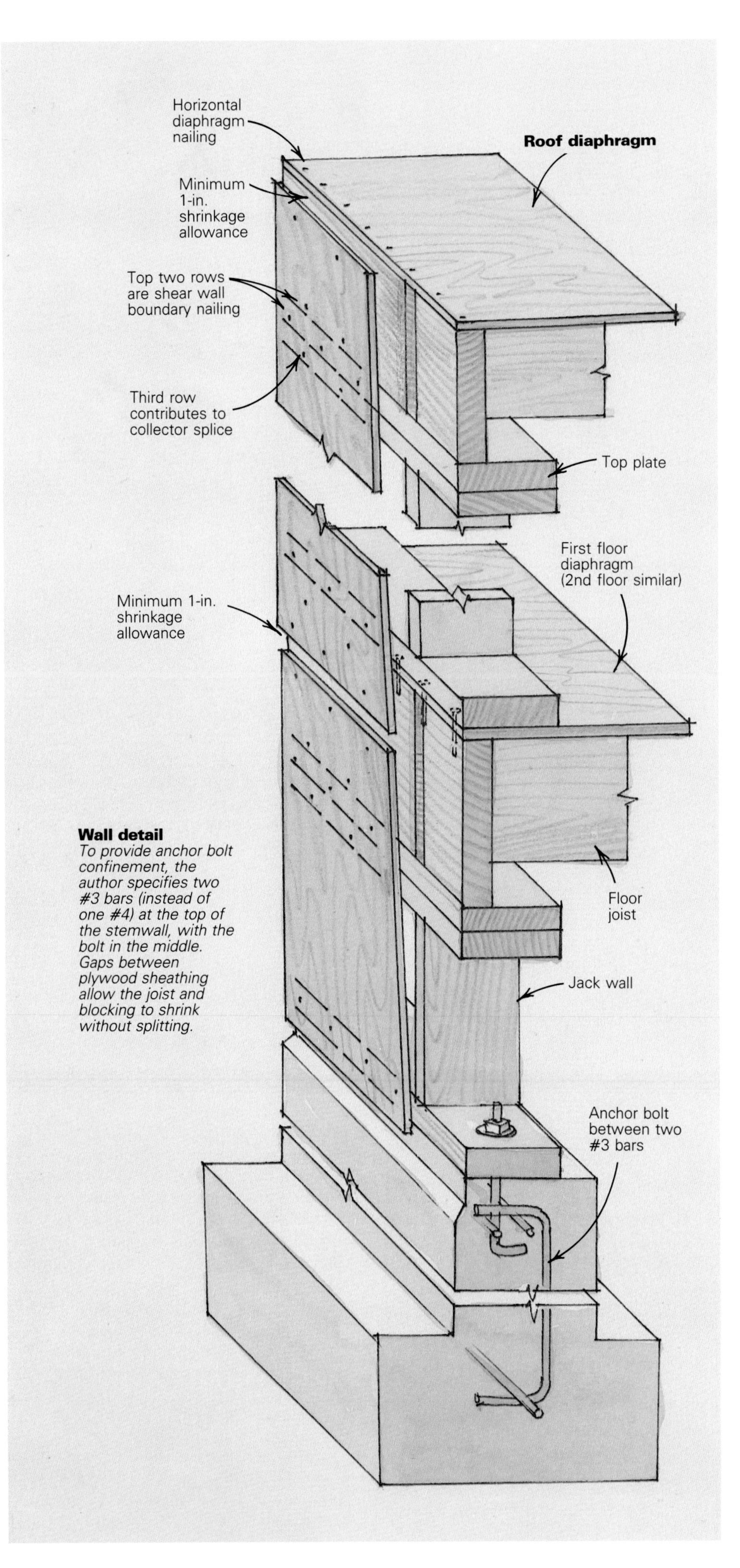

Wall detail
To provide anchor bolt confinement, the author specifies two #3 bars (instead of one #4) at the top of the stemwall, with the bolt in the middle. Gaps between plywood sheathing allow the joist and blocking to shrink without splitting.

correctly nailed plywood panels. Overdriven nails are typically installed by careless nailgun operators. If the gun sets the nails erratically, back off on the pressure, let them stand a little proud and drive them flush by hand.

In our tests, stapled plywood shear walls performed pretty well, but they weren't as strong as the nailed walls. Staples, being cold-worked, may be susceptible to brittle-failure, and being thin, subject to corrosion. This is an unhappy combination. So waterproofing the wall is more important when using staples than when using nails. Overdriven staples reduced strength, but not so badly as overdriven nails. Drywall, great for fireproofing and finish, should not be used for resisting earthquakes. As the walls flex, the nails just excavate little slots in the drywall. Bugle-head drywall screws, annular-shank or threaded nails, or regular screws are cold-worked, so they can't stand the repeated reversal and extreme deformation that common nails can. Don't use them for shear walls, unless you're certain they are annealed to the performance level of a common nail.

Lag bolts are fine for connections where loads are concentrated. But to work properly, the hole has to be drilled twice—once for shank and once for the threads—then lubricated with paraffin wax and the bolt turned in. No hammering allowed.

Another fastener that's subject to improper installation is the anchor bolt. It carries shear loads from the mudsill into the footing, or would if it weren't too close to the edge of the footing or outside the line of any rebar that might prevent the concrete from spalling during a quake. An equally useless method for installing anchor bolts is to stab them into the concrete after the pour has initially set—a disgusting practice that guarantees the bolt shank will be "anchored" in a cone of laitance (weak and crumbly concrete). Prior to the pour, anchor bolts should be wired inside a double run of rebar at the top of the stemwall (drawing left). The bolts' threads should extend far enough above the finished level of the concrete to accommodate the mudsill, the washer and the nut.

Moisture and shrinkage—Don't expect anything to work structurally when you really need it if it's decayed or being digested by termites. Make sure you've protected your work. Decay is the more insidious of the two. At the point when a specialist can only marginally detect decay under a powerful microscope, *80%* of the wood's shock resistance has vanished.

Strap ties used as tiedowns between floors will buckle as the joists dry, unless they're installed after the joists have shrunk. For the same reason, bolted tiedowns need to have their nuts tightened just before the walls are closed in. □

Ralph Gareth Gray is an architect and structural engineer living in Berkeley, Calif. He has designed wood-frame buildings for more than 30 years and served on code advisory committees for the Structural Engineers' Association of California and the American Institute of Architects.

Building Coffered Ceilings

Three framing methods

Editor's note: The following three projects hardly look the same, but they share one detail: a coffered ceiling. A coffer is characterized by sunken panels (they're usually square or octagonal) that decorate a ceiling or a vault. Though the term is generally associated with multiple panels, a proper coffer can have a single panel. The technique is thought to derive from the visual effect created in buildings where heavy ceiling beams crossed one another, and it has been used structurally and decoratively for buildings as dissimilar as neo-classical churches and the Washington, D. C., subway system.

Don Dunkley frames coffers into custom homes, typically by creating one big recessed panel. Greg Lawrence used coffering to conceal glulam beams. And Jay Thomsen used crisscrossed 1x wood strips to create the effect of sunken panels over the surface of a vaulted ceiling. *—Mark Feirer, editor of* Fine Homebuilding.

One big coffer. **Soffits girdle this bedroom to support angled coffer framing. A single, recessed ceiling coffer is the result. Photo by Scot Zimmerman.**

Single Coffer

by Don Dunkley

Among the most common ceiling details I run into when framing custom homes is the coffered ceiling. Though the term coffer encompasses a range of ceiling treatments, around here we use it to refer to a ceiling with a perimeter soffit having a sloped inner face that rises to a flat ceiling (photo left). The detail is usually found in bedrooms and dens.

The first coffers I built were usually sloped to match the roof and fastened directly to the roof framing. There was no soffit; the sloped portion of the coffer simply died into the surrounding wall. I used this method routinely for a few years—until I realized its limitations. For one thing, linking the roof to the framing of the ceiling limited the angle of the coffering to that of the roof (unless a very steep pitch was used on the main roof). Also, there was a limit to the amount of insulation that could be put into the perimeter of the coffered ceiling. Adding a soffit to the coffering solves these problems.

The soffit encircles the room and is framed so that its underside is level with the top plate. The soffit usually extends 1 ft. to 2 ft. away from the walls and offers several advantages. Framing is simplified, the pitch of the coffer can be any angle, there's plenty of room for insulation, and the flat ceiling surrounding the room can be embellished with can lights and crown molding.

The layout and the pitch of the coffer are usually found on the floor plan or the electrical plan. But before I start framing, I usually confer with the builder or the home owner to finalize the actual size of the soffit, the pitch of the coffer and the height of both the main ceiling and the soffit. Once these dimensions have been confirmed, the framing can usually be completed in a few hours.

From *Fine Homebuilding* magazine (June 1992) 75:36-42

Traditional Coffer

by Greg Lawrence

In the course of a recent remodeling project, we removed the roof from a 1,200-sq. ft. house and built a second-story addition in its place. We had to demolish the vaulted ceiling of the existing living room to make space for the new rooms above. To support those new rooms, we installed several glulam beams parallel to the exterior wall; the photo at right and the drawing below show how we coffered the ceiling to conceal the glulams.

First we wrapped each glulam on three sides with 1x Douglas fir, detailing the edges with a roundover beading bit and a router. Then we built intersecting false beams with 2x6 blocks (ripped to match the width of the glulams) and more fir. Finally, we trimmed the ceiling with crown molding. Where the molding returned off the window head casing, a striking horned cornice was created.

The resulting coffered ceiling adds a stately look to the room and nicely complements the window muntins. □

Greg Lawrence is the owner of Green River Construction in Sebastopol, Calif. Photo by the author.

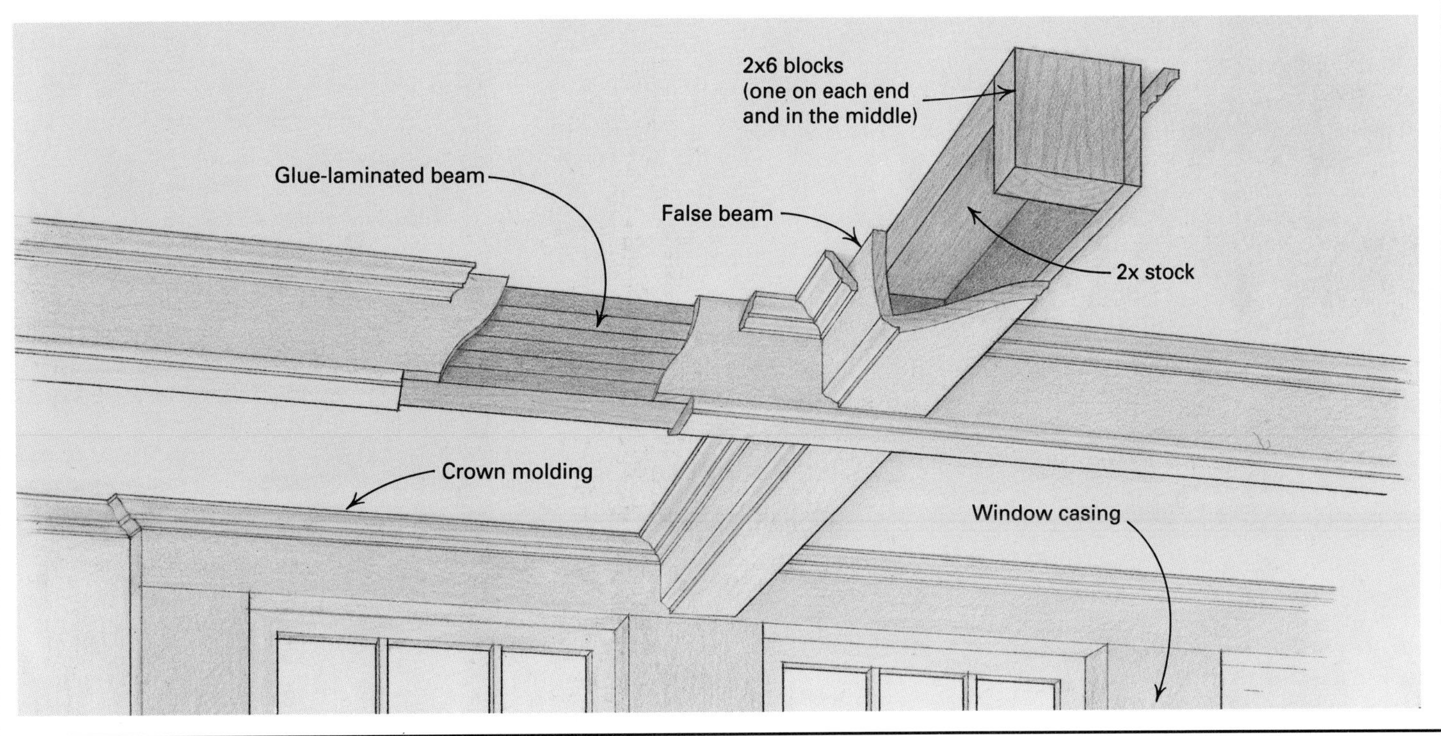

It starts with the soffit—One of the big advantages of the coffering technique I've adopted is that the coffer framing can be done before the roof is constructed. That gives us plenty of room in which to work. The first step is to lay out the location of the doubled joists, sometimes called carrier joists, that form the outer edge of the soffit (drawing, p. 66). The locations are marked on the top plates of the surrounding wall, the carrier joists are oversized because they support both the coffer framing and the soffit framing—we usually use 2x10s or 2x12s, depending on the size of the room. It's important to build this part of the framing (we call it a carrier box) straight and square. Otherwise, the rest of the coffer will be a bear to build, not to mention what the finish carpenter will say about you when he hangs the crown molding. Nail off all the carrier joists very well because green lumber, while drying, will try to go places you don't want it to visit; three nails spread the width of the boards on 16-in. centers will suffice. Of course, in order to build a good, square carrier box, the surrounding wall framing had best be on the money—a square box in an out-of-square room will endow the soffit with a noticeable deviation in width.

To install the carrier box, start by spanning the room (usually, but not always, the shortest dimension) with doubled carrier joists. Once these have been cut and nailed in place, string a dry line across each pair and brace them straight with a temporary 2x4 "finger." Nail the finger to the carrier, push the carrier into line, then nail off the finger to the underside of the top plate. This will hold the carrier in place until the framing is complete (top right photo, p. 67). After

Drawings: Vince Babak

A topless hip. **Once the soffit is in place, framing for the coffer itself is like a hip roof with the top removed. Pressure blocks are nailed between framing members on either side of the doubled carrier box; the blocks prevent the framing from twisting as it dries.**

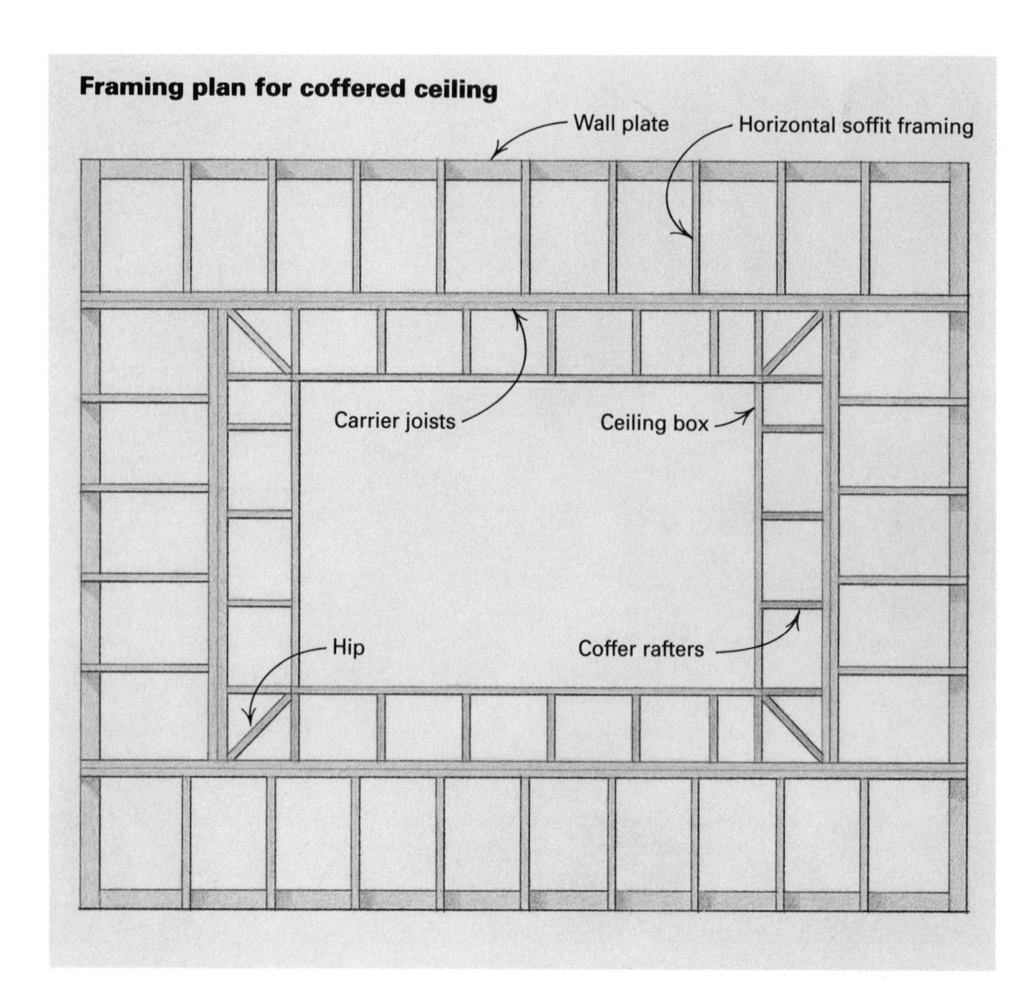

lining the first two pairs of carrier joists, measure and hang (we use joist hangers) the second two pairs between them. These carrier joists should be lined and braced as well.

With the carrier box in place, you're ready to lay out the locations of soffit joists on the top plate. We use 2x4s 16 in. o. c. for these joists, running them perpendicular to all four pairs of carrier joists (drawing left). The soffit joists should tie into the rafters at the exterior wall plates (a code requirement in these parts), so lay the rafters out ahead of time.

As we toenail the soffit joists to the plate with 8d nails, we secure pressure blocks in every other bay (photo above). A pressure block fits snugly between the ends of the joists to prevent them from twisting as the joists dry. Nail a 1x4 to the top of the joists that are toenailed to the plate, running it the length of the wall, and secure it with a pair of 8d nails at every joist. Called a catwalk around here, the 1x4 is required by code and helps to prevent twisting at the wall end. It should be located as close as possible to the intersection of rafters and joists.

One last check for clearance—With the soffit framing in place, you're ready for the angled coffer framing—but not before one last check of the specs. If the coffer is at a steeper pitch than the roof framing to follow, now's the time to make sure that the coffer framing won't interfere with the rafters. If someone changes the pitch of the

Framing the ceiling. **A ceiling box with mitered corners (photo above) forms the perimeter of the ceiling. A short hip rafter with beveled plumb cuts at top and bottom connects the corners of the ceiling box to the doubled carrier box.**

Helping fingers. **Pieces of scrap stock, called fingers, should be nailed between the carrier joists and the surrounding wall framing. They prevent the joists from bowing as the soffit framing is installed. Later on, the fingers will be removed.**

Blocking the rafters. **With the framing complete, Dunkley works his way around the ceiling to install any last pressure blocks that might be required.**

roof from what's on the plans, the angle and the height of the coffer should be recalculated—a quick double-check now can avoid major problems later when the roof gets framed.

To check this, measure the run from the inside of the exterior plate (in most cases, this is where the bottom edge of the rafter will start its incline) to the inside edge of the carrier box and add this figure to the run of the coffer rafter. This gives the overall run, and by plugging this into a calculator (I use a Construction Master II) and entering the pitch of the roof, you'll end up with the height of the roof rafter's bottom edge. When 6 in. is added to account for the thickness of the ceiling framing, you'll know if the coffer will collide with the rafters. If it will, lower the pitch of the coffer.

If the ceiling height hasn't been given on the framing plans, check a section detail (if there is one). A decent set of plans usually carries all this information, but not all plans are created equal. If the plans have left this information out, you'll have to calculate the height of the coffer based on the run and pitch of the coffer rafters.

The coffer layout—The coffer layout is no mystery; think of it simply as a hip roof with the top cut off (photo facing page). At each corner there will be two common rafters and a hip rafter; the areas between corners will be filled with common rafters. After laying out a common-rafter pattern, we cut as many rafters as we'll need. Mark the locations on the carrier joists of all eight commons that form the coffer corners, then pick one corner and work your way around the box, installing the fill rafters. These are usually 16 in. o. c., but 2 ft. o. c. is fine if the coffer is small. We use either 2x4 or 2x6 stock—in general, we use what we have most of. Of course, an unusually long span might call for larger stock.

Armed with the rise and run of the coffer rafters, you can figure them for length (for more on rafter framing, see *FHB* #10, pp. 62-69). There's no need to figure in a shortening allowance, though. When the length is known, we cut one pattern and then whack out the quantity needed. If we're building more than one coffer of the same size, the second set of rafters can also be cut now.

The coffer framing—After the rafters are cut (but before installing them), we build the ceiling box at the top of the coffer, which is similar to the carrier box that forms the soffit. The difference is that the ceiling box is smaller (by the run of the coffer rafters), and the framing is not doubled up. We usually frame it on the deck from 2x6 stock, then lift it into approximate position, using temporary legs to hold it up; these legs will rest on the floor. The frame should be square; carefully cut rafters will keep it straight.

Once the ceiling box is up, install a pair of common rafters at each corner to hold the box in place. Toenail the rafters top and bottom, then install the rest of the commons, adding pressure blocks to prevent the rafters from twisting later (bottom photo, p. 67). When installing the rafters, make sure that they're not bowing the ceiling box, trim them if necessary.

When the commons are in, cut the hips to finish off the corners (top left photo, p. 67). The hips will have double cheek cuts on both ends; the cuts can be measured in place or calculated. When installing the hips, fit them in so that the drywall will follow the plane of the rafters into the center of the hip. A 6-ft. length of 1x4 makes a good straightedge to guide the hip placement. Fill in any jack rafters, if needed.

The ceiling framing is simple: Just add joists inside the ceiling box and fill in between with pressure blocks (bottom photo, p. 67). We use 2x4s laid flat to provide backing for the ceiling drywall along the length of the ceiling box. A strongback can be run down the center of the joist span to prevent the joists form sagging.

Variations—There are several variations to our coffer-framing techniques. One way to install the ceiling box is to eliminate the temporary legs and install eight common rafters at the corners of the soffit. Then lift the ceiling box up past the commons until the bottom edge is flush with the bottom of the rafters. The pressure of the commons will hold it until everything's nailed off.

Another approach is to nail the ceiling frame to the commons one board at a time, eliminating the need for help in positioning the unit. This box is supported by the hip rafters. It can withstand quite a load as long as the lower ceiling box is well braced with the ceiling-joist fingers.

Crown molding—If crown molding is desired at the top of the coffer ceiling, the ceiling joists are placed on top of the ceiling box, allowing a 5½-in. recess for the molding. To blend the bottom of the coffer rafters into the inside edge of the recess, we rip the bottom edge of the ceiling box to match the rafter slope, providing a smooth transition. If the rip reduces the width of the stock too much, cut the commons with a notch to accommodate the ceiling box so that they will blend into the inside edge (the box will be oversized by 3 in. to make up for the notch). □

Don Dunkley is a framing contractor in Cool, Calif. Photos by Charles Miller except where noted.

Applied Coffer

by Jay Thomsen

Usually an addition is built to reflect the design of the main house. A recent project of ours, however, showed that the opposite can also be true: Eventually, the Mrachek's house will be remodeled to reflect the addition.

As vice-president of the Handel and Haydn Society in Boston, Massachusetts, Bobbi Mrachek wanted a room in which to entertain large numbers of guests, usually for live performances of classical music. The design delivered by local architect John McConnell called for a room 40 ft. by 15 ft. 16 in., topped with a barrel vault. The look of the ceiling had to be bold, but not dark and depressing, and the surfaces had to reduce the echo effect of such a large space. Coffering would soak up the echoes. To do this without further complicating an already involved ceiling structure, we created a coffered effect with strips of 1x stock (photo facing page).

The shape revealed. **The project began with standard gable roof framing. Flush cross ties (photo above) secured with metal plates roughed out the shape of the barrel vault. Plywood gussets (photo below) provided the final shape. They were screwed to the rafters.**

Providing the structure—As builders, we had constructed barrel vaults before, but never one so big. Given the dimensions of the new room, we knew we had a serious project to contend with. Fortunately, articles by Gerry Copeland and Lamar Henderson (see pp. 71-77) helped quite a bit.

We framed the main roof of the addition as a gable with 2x10 rafters. Flush-framed collar ties (photo top right) secured with metal gusset plates completed the basic shape of the room. Plywood gussets would form the exact curve of the vault and provide a nail base for the finish ceiling (photo bottom right).

Calculating the radius of the plywood strips was done simply by drawing a layout on the floor to exact scale using a string, a nail and a pencil. Sheets of ¾-in. CDX plywood were then laid out on top of the curve, and the radius was redrawn on top of the sheets. Three stacks of templates and several jigsaw blades later, we were ready to begin installation.

A working solution—The height of the ceiling (18 ft. 6 in.) became a factor in our planning at this point, not so much for safety reasons but for convenience: To hand each of the 1,350 pieces of the ceiling up a long ladder from below would have taken too much time. So we constructed a temporary second floor within the room to serve as a work platform—we'd need only a short stepladder to reach the highest point of the ceiling. A space 18 in. wide, running the length of the room, was left along each side of the platform to provide access for hoses, cords, passing up stock and even to dangle our legs through when working at the springline of the vault (the springline is the point of the ceiling where the curve first leaves the vertical plane of the wall).

The finish ceiling—After screwing the plywood gussets to the rafters and touching up the resulting curves with a belt sander, it was time to start nailing up the 1x6 finish ceiling. The Mracheks didn't want to see the V-groove that would characterize the seams of conventional T&G stock, so custom stock was milled from clear select pine. Every piece was prestained on both sides and both edges—twice—before installation and was sealed as soon as it went up.

We started the first piece of 1x6 at the centerline of the ceiling. Succeeding boards were then brought down either side toward the springlines. This allowed two crews of two men to work at the same time. We used nailers and 6d finish nails to secure the boards, toenailing most connections to avoid the incredible amount of time it would have taken to fill exposed nail holes with wood putty.

The coffer emerges—The main ribs of the coffering were to be layers of ½-in. pine applied in descending order of width (drawing, p. 70). Intersecting ribs would overlap each other. Our hopes were that the layering of each member

A vault of considerable size. This room addition was designed to house live performances of classical music. The lofty, coffered ceiling was created with built-up layers of 1x stock that were painstakingly screwed and nailed into place. Photo by James Shanley.

Adding up the ribs. **Stepped layers of 1x stock in various widths form the ribs of the coffer. Transverse ribs are slightly wider and one layer thicker than lengthwise ribs.**

Three against one. **The pliable nature of pine 1x stock allowed each layer of the ribbing to follow the vault of the ceiling. Nevertheless, it took a lot of work to press each strip into place.**

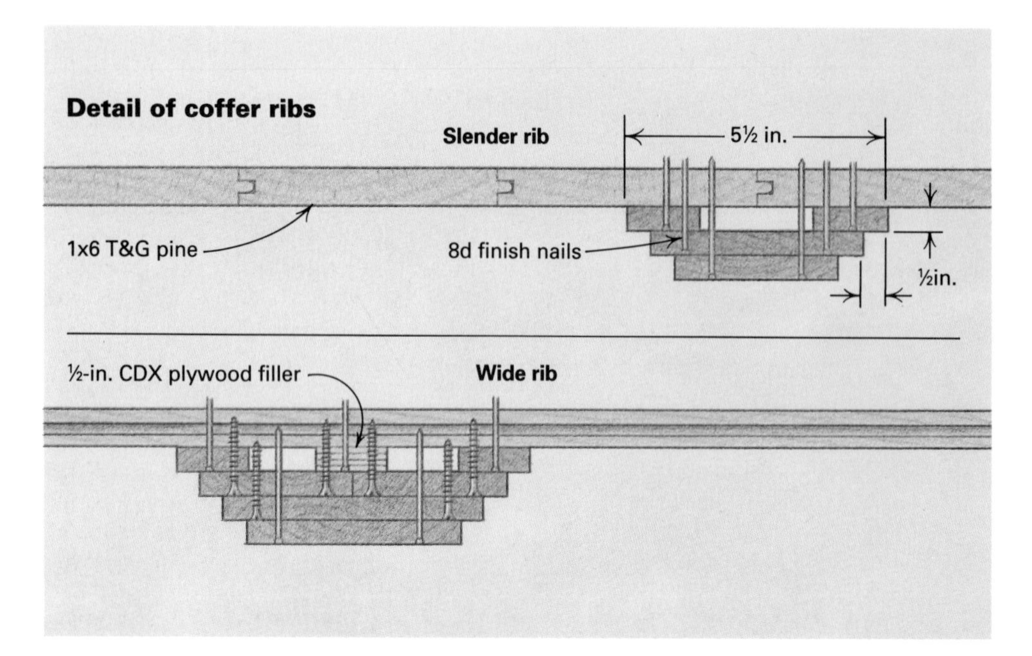

would create enough shadow lines to stand out from the background 1x6s and make the ribs seem deeper than they were (photo above). The heavier ribs that line up with the columns in the wall below are 7½ in. wide and one layer thicker than the smaller, intersecting ribs, which are 5½ in. wide. The larger ribs helped to break the ceiling into sections that were easily subdivided into smaller, coffered squares.

Fortunately, ½-in. pine conformed to the radius of the ceiling, although it took three carpenters to bend and fasten each member in place (photo left). Each layer, except the final one, was screwed in place with 1⅝-in. drywall screws; subsequent layers hid the screw heads. The final layer of each rib was secured with 8d finish nails. In some cases, especially near joints, we needed better holding power, so we used 1⅝-in. trim screws instead of nails. All pieces were installed with butt joints; we were afraid that the beveled ends of scarf joints might slide past each other as the layers were fastened into place.

Given the repetitive but very precise nature of the coffering, it was important to come up with an accurate, easy method for laying out the location of each member. Ribs were kept parallel to one another by constantly checking measurements off the end walls and by our consistent use of spacer sticks cut to the desired distance between ribs.

The layout of the ribs running perpendicular to the main ribs (parallel with the springline) was kept in line with a 10 ft. long (and very flexible) layout stick, marked with the desired rib locations. As long as the end of the layout stick was butted to the springline, and the length of the stick was snug along the ceiling curve, the layout stayed very consistent. The spacing for all ribs is approximately 2 ft. o. c.

All in all, the vaulted ceiling consumed approximately 640 man hours. If we had a similar ceiling to do again, we could probably cut about 50 man hours from the process.

All carpentry work on this job was done by carpenters Charles Desserres (lead), Brian McCune, Don Baker, Steve Harris and Mark Roberto of I. M. Hamrin Builders, Milton, Massachusetts. □

Jay Thomsen is a remodeling contractor in Milton, Mass. Photos by Charles Desserres except where noted.

Building Barrel Vaults

Two ways to ease construction with modern materials

Editor's note: Presented on the following pages are two very different approaches to building a residential barrel vault with modern materials. A vault is an arched ceiling or roof, and its precedent as an architectural form was probably set by the cave dwellings of prehistoric man. The Egyptians used bricks to build the first manmade vaults about 4,000 years ago. Engineering the vaulted form has been a recurring theme in architectural history ever since.

Masonry vaults exert significant outward thrust on their supporting walls, and for centuries the efforts to resist that thrust led to innovative construction techniques—groin vaults, pointed arches, flying buttreses. But toward the end of the 19th century, with the advent of lightweight steel frames for construction, outward thrust was becoming less of an issue in vaulted buildings. In the 20th century, the steel reinforced-concrete shell was developed. This allowed the construction of a vault that exerts no lateral thrust and can be supported just on the ends as if it were a beam. Today the engineering and construction of vaults continues to evolve in response to new building materials.

The barrel vault (also called tunnel or wagon vault) is semicircular in section and is the simplest form of vault. —Kevin Ireton

Trusses and Plywood Gussets

by Gerry Copeland

When I design a house I don't usually start with a dramatic geometric shape already in mind. Most of my designs evolve from site determinants, function, client preferences and budget. However, recently I built a house on speculation, and I wanted it to stand out from its conservative competition in suburban Spokane.

I had recently visited a small Episcopal church in Charleston, S. C. The nave of the church was dominated by a spacious great room with a voluptuous barrel-vault ceiling. This experience, along with a hankering to do some curved detail work, set me in a determined direction. Hopping on the postmodernist bandwagon, I designed a traditional gable-roofed house around a great room, with a vaulted ceiling front to back, a large Palladian window and some whimsical columns and capitals.

Design—I made the vault 16 ft. wide because that seemed to be the smallest semicircular shape that was still a functional space. Because the rear of the great room was to have a balconied loft, the curved ceiling needed to be an average usable height. I considered an 8-ft. radius to be the minimum for this.

For visual impact I wanted the space to be open from the front of the house to the back. A 16-ft. wide Palladian window at the front of the house and a 5-ft. square window at the back (selected because of cost restraints) provide dramatic lighting.

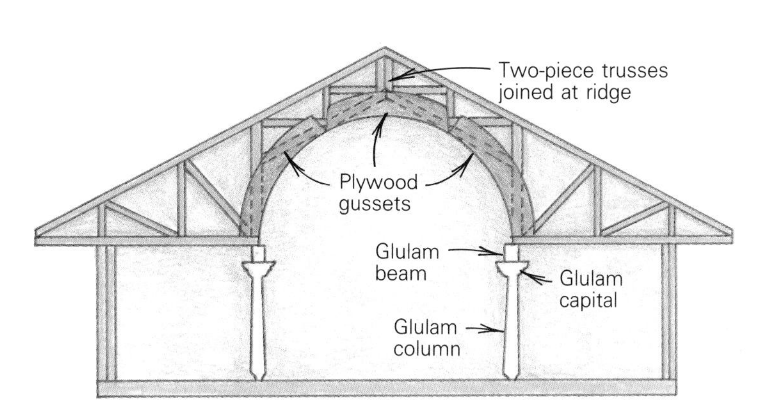

The basic structure of the house seemed obvious once I laid out the central vaulted space, drew the roof pitch I wanted (6½ on 12) and added supporting columns at the bearing points. By providing a minimum of 18 in. for insulation and ventilation at the narrowest point betweeen the vaulted ceiling and the roof, the roof-truss configuration emerged. A 44-ft. long house-width truss with a bite out of the middle would have been too flimsy to transport and erect. The truss fabricator suggested dividing the truss into two at the ridge line, with two parallel-chord sections cantilevered over the vaulted space to meet in the middle (drawing left).

The design of the rest of the house fell into place quite easily, though with some compromises to the plan in order to keep the central vaulted space as a strong visual element. It was, for example, important to keep the vaulted space uncluttered by inter-

To create the barrel-vault shape, curved plywood gussets were stapled to the roof trusses. The ceiling was finished with 1x6 T&G pine nailed to the plywood gussets. The boards had to be hand-nailed because pneumatic tools wouldn't draw them up tightly against the framing.

secting wall planes. Being a builder as well as an architect, I was eager to work out the details and start framing.

Building the barrel—After all the support columns, beams and walls were in place, the trusses were set by crane onto the structure, one bundle stacked flat on top of the framing at each end of the house. Then my crew and I rolled each truss into its upright position, first one half of the pair, then the other. The cantilevered top sections straightened up nicely once they were pushed together, aligned and nailed. We stapled ½-in. plywood gussets across the adjoining webs at the ridge to tie the two trusses together. After all the trusses were erected, double-checked for alignment and nailed down, we cross-braced them according to the truss manufacturer's instructions, ran solid blocking between them over the exterior walls, and then sheathed the roof with ⅝-in. CDX plywood.

Following the wall sheathing and shingling, we cut the curved gussets that would be attached to the inside surfaces of the trusses to form the vault. After much deliberation over which material to use, we decided on ⅝-in. CDX plywood. In retrospect, a higher-quality ¾-in. plywood would have been a better choice for greater stiffness and a thicker nail base. In order to find the most cost-effective way of cutting the plywood, I spent an evening laying out the curves to scale on paper. I ended up with a cutting solution that yielded four curved gussets, 8 in. high and 8 ft. wide, per sheet.

To scribe curves on the plywood, we used an 8-ft. long wire, wrapped around a pencil on one end and around one of five nails on the other. The five nails were driven into the floor 8 in. apart and represented the centers of each arc. This allowed us to draw them quickly. Then we cut the gussets with an orbital-action jigsaw. Altogether we cut 17 sheets of plywood for a total of 66 gussets.

I'm not sure whether it was by luck or by intuition, but any three of these gussets together, point to point, made a half-circle that was exactly 16 ft. wide. So by lining up the ends of the gussets with the opening in the trusses, we could assure perfect alignment. The gussets were stapled to the trusses with 2-in. sheathing staples. When all the gussets were in place, we sighted down the 42-ft. length of the barrel and saw that the alignment was perfect. At this point in the framing, someone looked up at the exposed framing and said, "This is what the ribs of the whale must have looked like to Jonah" (photo above left).

Siding the ceiling—We used 1x6 T&G pine, prestained with two coats of semi-transparent stain, to finish the ceiling (photo above right). Keeping a straight line for 42 ft. with only a slender ⅝-in. plywood edge to nail to would not be easy, so we eliminated all the crooked pieces, as well as those pieces with loose knots. Because of the barrel-vault shape and the T&G connection, board-to-board nailing into the edge of ⅝-in. plywood seemed strong enough. But if I were to do it again, I'd use ¾-in. plywood and construction adhesive at each rib. In order to get a straight start, we chalked a line at the base of the vault, along the trusses. The first two or three courses could be nailed directly into the bottom of each truss. We decided to use 6d finish nails driven with hammers because our pneumatic nailers wouldn't draw the boards up tight.

We quickly wished that some benefactor had donated perfectly straight and clear T&G cedar for the entire job. Every two or three rows, we sighted down the barrel and compensated for waviness by prying away from a gusset the boards that were bowed inward. All butt joints were beveled at a 45° angle. These joints looked good a year later but would have been better still had the lumber been perfectly dry.

The 16-ft. Palladian window was made for us by a local cabinetmaker and custom window fabricator. Most window manufacturers limit the size of their efforts to windows 8 ft. or less in diameter. Because of this window's size, extra thick muntins, 3-in. by 6-in., were used to withstand lateral wind load. It was built and shipped to the site in one piece, and installing it was a struggle for three of us. The double-insulated glazing was installed by window-glass fabricators after the frame was in place.

Glulam columns—Along the front half of the barrel vault, the roof trusses bear on 6-in. by 16-in. glulam beams, supported by whimsical columns cut in a classical profile (photo facing page). The columns were cut from 5⅛-in. by 12-in. standard architectural-quality glulams made up of laminated 2x6s. Because they were to be painted, minor blemishes and construction bruises could be filled, sanded and finished upon completion. We made the 8-ft. long taper cuts on each side of the columns by run-

The front half of the barrel vault is supported by a pair of 6-in. by 16-in. glulam beams, which in turn are held up by tapered glulam columns.

Photo: Joe Nuess

ning the pieces through a large bandsaw. Even with a new 1-in. wide blade we could barely cut straight enough to enable a 6-in. hand-held power plane to smooth out the irregularities. All four edges along the column's length were finished with a router and a ¾-in. beading bit, starting 6 in. down from the top and stopping 10 in. up from the bottom.

The column capitals with the tight radius cuts were done easily on the same large bandsaw using a ⅜-in. blade. These short pieces of glulam were easy to handle. We made the curving cuts so cleanly that only a minimum of sanding was necessary to finish. The edges were dressed with a ⅜-in. roundover bit. Then we attached the capitals to the columns by drilling down from the top and fastening them together with two ⅜-in. by 12-in. lag screws. The bottoms of the capitals were notched 1 in. to sit over the tops of the columns.

My barrel-vaulted spec house definitely stood out from its conservative competition, but was on the market for an agonizing two years before it eventually sold. □

Gerry Copeland is an architect/builder in Spokane, Washington. Photos by the author except where noted.

Longitudinal Wood I-Beams

by Lamar Henderson

Drawing a simple arc on an elevation can change the entire approach to the design and construction of a project. At least that's what happened with the Gottlieb/Davis remodel. Dan Gottlieb and Peggy Davis had originally wanted a kitchen remodel and greenhouse addition to their house in Palo Alto, California. But after analyzing their programmatic needs, the existing residence and site, as well as property values in the neighborhood, we agreed that a more extensive remodel was appropriate.

The general massing of the house took on the appearance of two row houses, each reflecting a different living zone. One became the private zone: the bedroom area with a pitched roof. The other was the public zone: living, dining, kitchen, sunroom, second-story library and a bridge to the other side. We jokingly referred to the public space as the grand hall. In the search to find a dramatic roof for the grand hall, I sketched a barrel-vault ceiling and changed the course of the project.

I submitted the barrel-vault design to Dan and Peggy as one of three proposals. They were apprehensive, but at my suggestion, they decided to build a scale model, and once they were able to visualize the design, the barrel vault became the obvious choice.

We knew that the complexity of building the vault would either deter some contractors from bidding on the project or would result in highly inflated bids. So after a thorough discussion, Dan and Peggy and I decided to build the project ourselves.

Design dilemmas—Conceptually, the grand hall is a basilica with a nave and side aisles. The aisle on the south became the entry, stairs and sunroom. And the loads from the vault would flow through the wall to the foundation. The northern aisle, however, was problematic. Due to the constraints imposed by the floor plan, there were no walls that could take roof loads directly to the foundation. A typical strategy of using a curved truss at intervals to shape the vault would require a very large and heavy beam to carry the loads to the end walls. I had to find an alternative solution.

The elevations of the two end walls dictate the shape of the roof and interior space, so it seemed logical that the structure should span the length of the hall with loads flowing down through these end walls. The problem was to find a structural member that could span 36 ft. and yet be light enough to carry by hand. A glulam beam, for example, would have been too heavy.

The metaphor that the vault was, in effect, a floor of parallel joists in the shape of a circle, suggested that some form of lightweight wood truss could span the distance. After considering the cost and weight of various engineered components, I decided on wood I-beams with solid-lumber flanges and oriented strand-board webs purchased from Structural Development, Inc. (SDI, P.O. Box 947, Los Gatos, Calif. 95031).

Attaching the I-beams to the end walls was no problem because manufacturers of metal connectors for wood construction have developed special hangers for I-beams. Developing a blocking system to tie the long I-beams together was more of a challenge. When I initially laid out the I-beams on the end wall, aesthetics ruled over good engineering. I called for the I-beams to be laid out radially so there would be a flat surface to which to attach the exterior roof sheathing as well as the interior drywall. This created a problem because I-beams, though strong when loaded vertically, are flimsy when loaded horizontally. I planned to solve this in part through the use of blocking, cut to the curve of the arc between adjacent flanges.

When the project passed through the Palo Alto building department, the plan checker asked for a more detailed analysis of the shell structure, especially the rotation of the beams in their weak axis and the behavior of the entire roof assembly. To find out how to do this, I contacted the American Plywood Assoc. (P. O. Box 11700, Tacoma, Wash. 98411). The call turned out to be the nadir of the project. One of their engineers said the structural system should be analyzed as a curved stress-skin panel. Not only would this be mathematically complicated, but fabricating such a panel in the field to achieve the assumed structural values would be very difficult because the plywood would have to be glued to the frame under pressure.

To eliminate the need for a shell analysis and deal with the problem of having the I-beams in the weak axis, we rotated all the I-beams in their vertical axes, thus maximizing their value as structural members. Now, the problem was to create an outside curve that would allow the roof diaphragm to work, and an inside curve would create the barrel-shape ceiling.

Two-by blocking could be cut to fit the shape of the curve and attached perpendicular to the I-beams with Simpson H4 metal anchors (Simpson Strong-Tie Co., Inc., P.O. Box 1568, San Leandro, Calif. 94577). This would make the top flanges of the I-beams rigid and create a curved surface to glue and nail the plywood. However, the bottom flange and portions of the I-beam could still move laterally. We figured the bottom flange could be made rigid by using some type of metal strapping as bridging.

Because of the way the I-beams were being installed, the bottoms

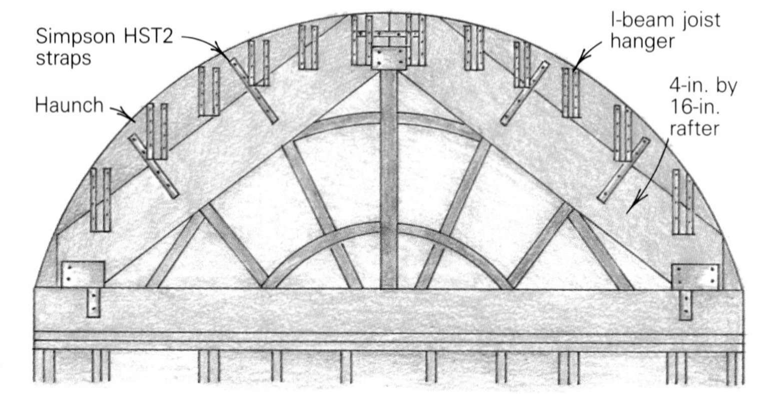

Drawings: Michael Mandarano

of the beams would not line up in a smooth arc and would not work as a nailing surface for the drywall ceiling. However, a quick review of the *Gypsum Construction Handbook* (United States Gypsum Co., 101 S. Wacker Dr., Chicago, Il. 60606-4385) showed a commercial system that would work for us. It employed cold-rolled steel channels and steel furring channels as a method for creating the interior vault.

After correcting the drawings, we resubmitted them to the building department for final approval. The building permit was issued and construction began.

Beaming up—The first order of business was to order the two curved glulams that would shape the end walls. We sent a drawing to a local glulam fabricator, and the bid came back at just under $6,000. Why were the glulams so expensive? Apparently, the diameter was less than the minimum allowed by the fabricating jigs, using ¾-in. stock. In order to make the glulams at such a small radius, the laminate had to be ⅜ in. thick. Material any thicker would have failed in bending.

The cost was too high. We finally solved the problem by using a pair of 4-in. by 16-in. timbers (select structural Douglas fir) as rafters, with separate pieces (called haunches) shaped to the curve of the vault and bolted on top of the rafters (drawing facing page).

The next step was to make a template of the end-wall elevation that we could use to lay out and fabricate the windows. We laid out four sheets of ½-in. plywood and struck a radius for the outside edge of the vault as well as for the head and sill of the windows. Next I drew the center 4x4 column, then the other 4x4 members dividing the window frames. I drew the width of the 4x16 rafter, locating the curve and the haunch. All metal connectors such as the straps for the haunch and the hangers for the I-beams were also located on the template.

We cut out the plywood template, including the window openings, with a portable jigsaw. After carefully marking the pieces of the template and setting them aside, we sent the window cut-outs to the window manufacturer to use as templates for the glass.

The floor space of the second-story library was big enough to lay out and prefabricate the end walls of the vault. We cut the rafters and their haunches with a 7¼-in. worm-drive circular saw, cutting partway through from each side. We found, however, that we had to have two saws: while one was in use, the other was cooling off.

After all the pieces were cut, the end wall was assembled, checked, bolt holes were drilled for the haunch straps and hangers were located for the I-beams. First we installed the center 4x4 post, followed by the left and right 4x16 rafters, the haunch, which was bolted down using Simpson HST2 straps, and the 4x4 frames for the windows. Finally, the hangers were nailed to the rafter/haunch assembly. As the end walls were being cut and installed, the blocking and shims were being mass-produced.

With the end walls in place, it was time to install the I-beams. We started with 18-in. deep I-beams, one on each side of the vault. The length was carefully measured on each side to verify that the end walls were square and plumb. We cut the I-beams and nailed web stiffeners at the end bearing points. (Web stiffeners are vertical pieces of 2x4s cut to fit between the upper and lower chords of each truss.) The 36-ft. beam weighed approximately 150 lb.; four people could easily maneuver it into place. The next beams were 16-in. deep, weighing about 144 lb. each. The eight remaining beams were 14-in. deep and weighed about 137 lb. each. We installed the curved blocking (16 in. o. c.) and shims as we went along in order to stabilize the I-beams (photo below). Then we crisscrossed the I-beams with metal strapping on the upper and lower flanges every 32 in. to stabilize the beams even further. At the suggestion of Kurt Anslinger from SDI, we used plumber's tape (the metal strapping that plumbers use to hang pipes) for this, which worked just fine.

According to the American Plywood Association, ½-in. plywood bent widthwise has a minimum bending radius of 6 ft., and that would work for this roof. We nailed down the sheathing with 8d ring-shank nails to the beams and to the blocking. Between the courses of blocking where the plywood edge had no support, we used H-shaped plywood clips to tie adjacent sheets together.

Before we started the interior finish work, we heeded another bit of advice from Anslinger and used duct tape to wrap the strapping where the crisscrosses touched each other. As the building moves, he had told us, the straps could rub against each other creating a bothersome noise that would be hard to fix once the ceiling was in place.

We vented the roof by drilling a series of 3-in. dia. vent holes, 18 in o. c. along the tops of all the I-beam webs. Then we installed three wind turbines on the roof and continuous soffit vents on the south side. Because there are no exterior soffits on the north side—where the addition joins the rest of the house—we added three eyebrow vents to the roof on that side. We insulated between the I-beams with 9-in. R-30 fiberglass batts.

Cold-rolled ribs and furring—Talking with a technical representative from the USG, I learned that to build a vault like this I could use 16-ga. cold-rolled steel channels for the ribs, bend them to the desired radius and then attach them to the structure. We decided to

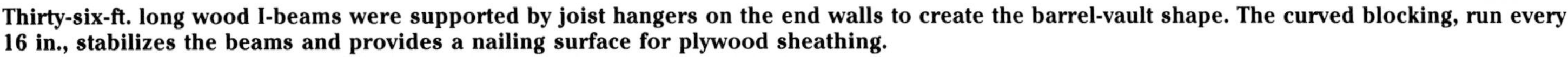

Thirty-six-ft. long wood I-beams were supported by joist hangers on the end walls to create the barrel-vault shape. The curved blocking, run every 16 in., stabilizes the beams and provides a nailing surface for plywood sheathing.

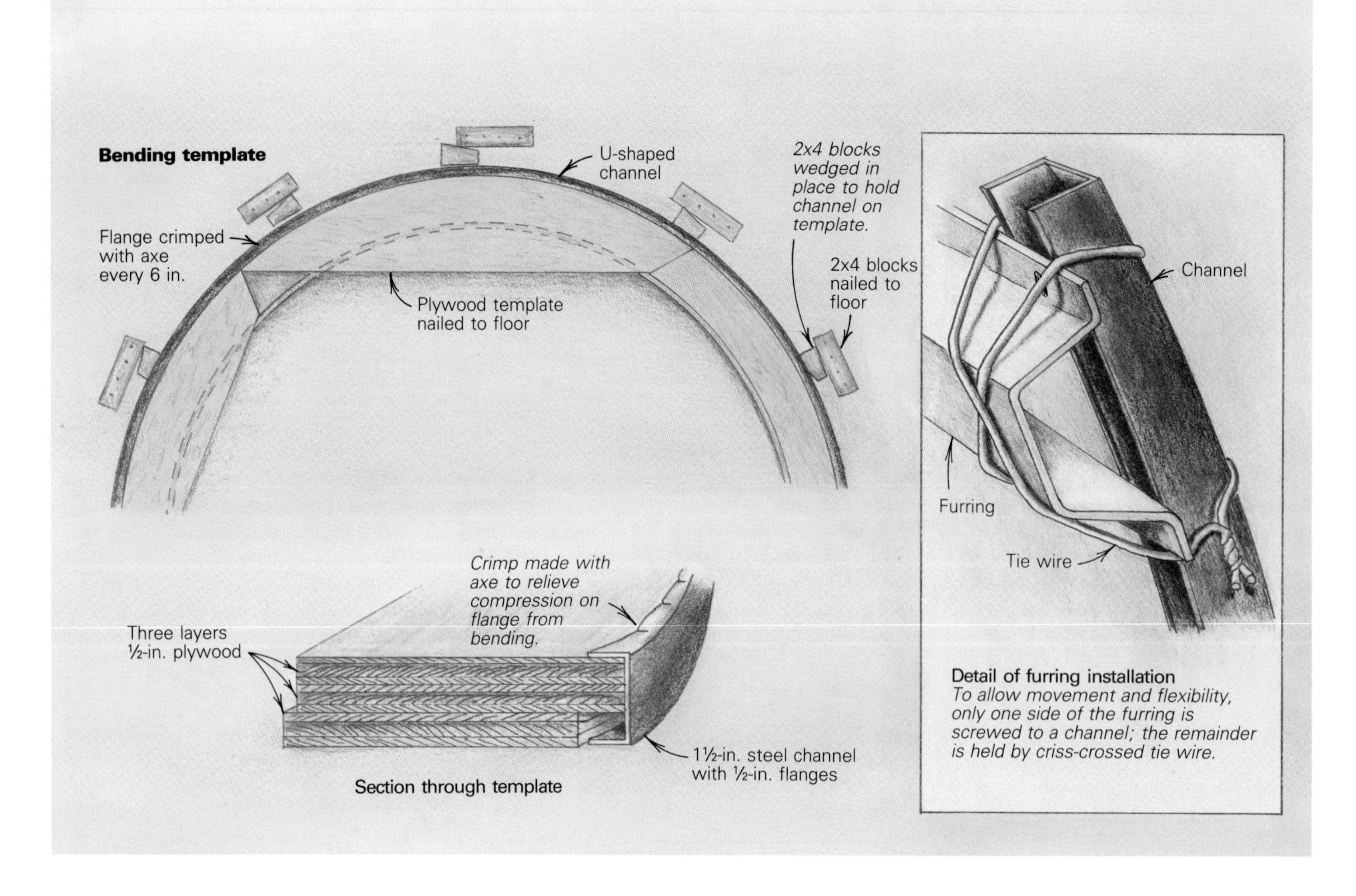

Detail of furring installation
To allow movement and flexibility, only one side of the furring is screwed to a channel; the remainder is held by criss-crossed tie wire.

use a U-shaped channel 1½ in. wide with ½-in. flanges, installed 24 in. o. c. Then we would screw and wire metal furring every 16 in. perpendicular to the ribs. The furring, made of 25-ga. galvanized steel, is hat-shaped in section.

To find out how to bend and install this system, I called the California Drywall/Lathing Apprenticeship and Training Trust in Hayward, California (23217 Kidder St., Hayward, Calif. 94545-1632). In conjunction with their training program, they have produced a series of film strips for training apprentices on all applications of drywall, plaster and lathing. They had two films on the sequence and installation of barrel vaults.

From the first film strip I learned that bending the U-channel and maintaining its structural integrity is critical. The channel should be bent with the flanges toward the inside of the vault. Since the flanges are in compression, the material can buckle and fail while being bent, thereby losing the smooth shape of the curve. The film strip showed a technique for bending the channel using a special bending device. It also demonstrated that you should screw only one side of the hat-shaped furring to the channel. Tie wire, crisscrossed and tightened, was used to secure the remainder of the furring (detail drawing above). This holds the furring flush with the cold-rolled channels, yet allows some movement and flexibility.

The second film strip showed the installation procedure for the gypsum board. A chart (taken from the *Gypsum Construction Handbook*) listed the minimum bending radii of dry-gypsum drywall by thickness. For our vault, two ¼-in. pieces of dry material could be used, or by moistening the back paper thoroughly prior to application, ½-in. gypsum board could be used. The moistening was necessary to allow the back paper and gypsum (in tension) and face paper and gypsum (in compression) to stretch and compress without failure. Of course, drywall screws had to be used to attach the gypsum board to the furring.

The axe-swing dance step—Now that I knew how the vault had to go together, I had to find the bending device. I called every rental place in the San Francisco Bay Area. No luck. Then, I called lathers and drywall contractors to find out if they would bend the cold-rolled channel. Again, no luck. Instead they wanted to bid the entire job. So, how could we bend the channel?

The solution to our problem slowly emerged. We built a full-size plywood template of the arch out of ½-in. plywood, with ½-in. plywood spacers underneath, and nailed it to the floor (drawing above). The U-channel fit over it with the bottom snug against the edge of the plywood. To hold the channel in place around the template, we nailed short 2x4s every 3 ft. around the outside of the arc, leaving enough room to drive wedges between them and the template. As the channel was bent around the template, we tapped a wedge into place. The length of the arc was just under 20 ft., and since the channel came in 20-ft. lengths, only one piece was needed per rib.

Unfortunately, the channel would straighten out whenever we removed the wedges. We tried dimpling the flange every 6 in. by hitting it with a cold chisel and a hammer. This took forever and didn't dimple the metal enough—it was hard on the template, too. We needed a faster way.

Dan happened to have an old axe in his tool box, and he decided to give it a try. By quickly hitting and denting the flange with the axe every 6 in., then turning the channel over and doing the same thing to the other side, we found that the channel would retain the shape of the arc of the template.

We had 20 ribs to bend, so we developed a rhythm for mass production. We called it the "axe-swing locomotion" dance. By rotating one's feet together from heel to toe and striking the flange with the ax, it was possible to develop a rhythmic pattern that created just the right swing motion to dent the flange uniformly. It took a great deal of hand/eye coordination and physical control on the swings to make it work, but agile dancers that we were, the job took no time at all.

The ribs were laid out at 2-ft. centers under the bottoms of the I-beams. The system was checked for roundness, plumb and parallel. The ends of each rib were securely anchored by screwing them to a metal stud run on top of the walls along both sides of the vault. At alternate I-beams, a ½-in. hole was drilled through the web, and the rib was tied to the I-beam with wire. The ribs were tied with a little slack in order to allow movement and to encourage the weight of the system to rest on the walls instead of hanging from the roof structure.

After the ribs were installed, we attached furring at 16 in. o. c. We continued with this procedure across the length of the roof, moving the scaffolding as necessary. Nine ft.

Inside the barrel vault, the drywall ceiling was screwed to a metal framework of arched ribs and straight furring hung from the bottom of the I-beams (above). Two layers of ¼-in. drywall were used to form the ceiling. The second layer was run perpendicular to the first. Residential drywall contractors shied away from the project, but a commercial crew that was between jobs happily took it on. They hung, taped and finished the job (below), including the barrel-vault ceiling, in three weeks.

Photo by Staff

from the end of each wall, at the top of the barrel vault, an electrical box was installed for ceiling fans.

Hanging the drywall—Finally, it was time to install the two layers of ¼-in. drywall on the barrel-vault ceiling. We ran the first layer parallel to the ribs (photo above) and the second layer perpendicular to the first. Soon after we started applying the drywall, our two laborers left the job to return to school. Because this was one of the least-pleasant tasks of the entire project, we decided it was time to call in a professional.

Most residential drywall installers weren't interested in working on the project because of the metal furring system, the ceiling height and the overall complexity of the project. However, we were lucky to find a commercial installer who needed to fill a gap in his work schedule to keep his crew employed. They hung, taped and textured the house in three weeks.

After they finished, we still had to cover the nuts and bolts where the straps held the haunch to the rafters. We found a local firm (San Francisco Victoriana, 2245 Palou Ave., San Francisco, Calif. 94124) that makes decorative trim plaster castings for Victorian houses and ordered 12 plaster keystones. They were hollow on the back and fit nicely over the exposed bolts (photo right).

It was hard to believe that the simple arc sketched on an exterior elevation 18 months before was now a three-dimensional reality.□

Lamar Henderson is an architect in Palo Alto, Calif. Photos by the author except where noted.

Header Tricks for Remodelers

Creative responses to unusual specs

by Roger Gwinnup

Between the Scylla of the homeowner's desires and the Charybdis of the house's structural needs, the course of the remodeler is often narrowly charted. When the request is to "take out that wall, but hide the header so it looks like nothing ever happened," the possibilities are usually at least two, but hopefully not more than two million. Given the vagaries of everyone who has ever worked on the house previously, the only way to proceed is to get out the old recipro-saw/scalpel and make an incision. Having done so several times, with instructions to "hide that header," I'd like to share some techniques that have worked for myself and my former partner Bill Pappas, now in Minnesota.

Heading off truss ends—I once built a small, gable-roof addition to a house that had pre-engineered roof trusses. The header had to run perpendicular, and adjacent to, the truss ends. I plumb cut the truss ends flush with the outside of the old framing (being careful not to disturb the truss plates) and removed enough plywood sheathing to allow access for nailing (drawing 1). Using a metal-cutting blade in my reciprocating saw, I cut the nails holding the truss to the wall and slid a joist hanger on the end of each truss. After installing posts at either end of the opening, I set the header and nailed the joist hangers to it. All that remained was to secure additional joist hangers to the other side of the new ceiling joists, remove the wall, and have a donut.

One advantage to this method, besides the concealed header, is that the existing wall supports everything until the new header is completely installed. Another advantage is that starting wall removal from the outside of a house leaves everything in place as a barrier between a habitable room and a construction zone. Depending on the situation, several wall components may be reusable, especially the insulation and the studs, if the nails have been cut from the top and bottom plates and the studs carefully twisted away from the drywall.

Balloon-frame header addition—I once worked on a two-story balloon-frame house with 2x6 walls where I needed to add a one-story addition with no visible header over the opening between them.

I put in a temporary wall inside to support the second floor; I then notched out the existing wall studs to receive the new header (drawing 2, facing page). Because the studs in a balloon frame extend in one continuous length from sill to roof, I notched the header 1¾ in. into the wall studs to better support the existing second floor. For this particular job I used two Micro=Lams to form the header. The advantage of this is that I can go to my supplier with the spans and load factors, and they will calculate the size I need. I secured the first Micro=Lam to the framing with 16d cement-coated sinkers, then attached the second to the first with more 16ds and Max Bond Adhesive (H. B. Fuller Co., Building Products Div., 315 S. Hicks Rd., Palatine, Il. 60067; 708-358-9555), though any good construction adhesive will do. I installed support posts at each end of the opening, and used twisted hanger straps to connect each existing joist to each stud and to the Micro=Lam. Then I removed the studs below the joists and hung the new ceiling joists.

Concealed header in an attic—I once removed part of a bearing wall in the middle of a house that supported the ceiling joists from each side of the house. In this case I went up in the attic with some 2x10s and laid them across the top of all the joists that were going to lose bottom support.

Then I installed twisted metal straps (drawing 2, facing page) to connect the header and the joists, went back downstairs, and removed some drywall and a stud from each end of the wall to be removed. After making sure that the posts themselves were supported below to the foundation, I slipped them in place and removed the rest of the wall. The only patching necessary here was to install 4½-in. wide strips of drywall along two walls and the ceiling, and to patch the flooring.

Removing ceiling sag—This trick involves the use of a strongback, but the principle would work for a header, too. I once worked on a motel that had two large dining rooms divided by a large hanging curtain. The curtain track was attached through the ceiling directly to the bot-

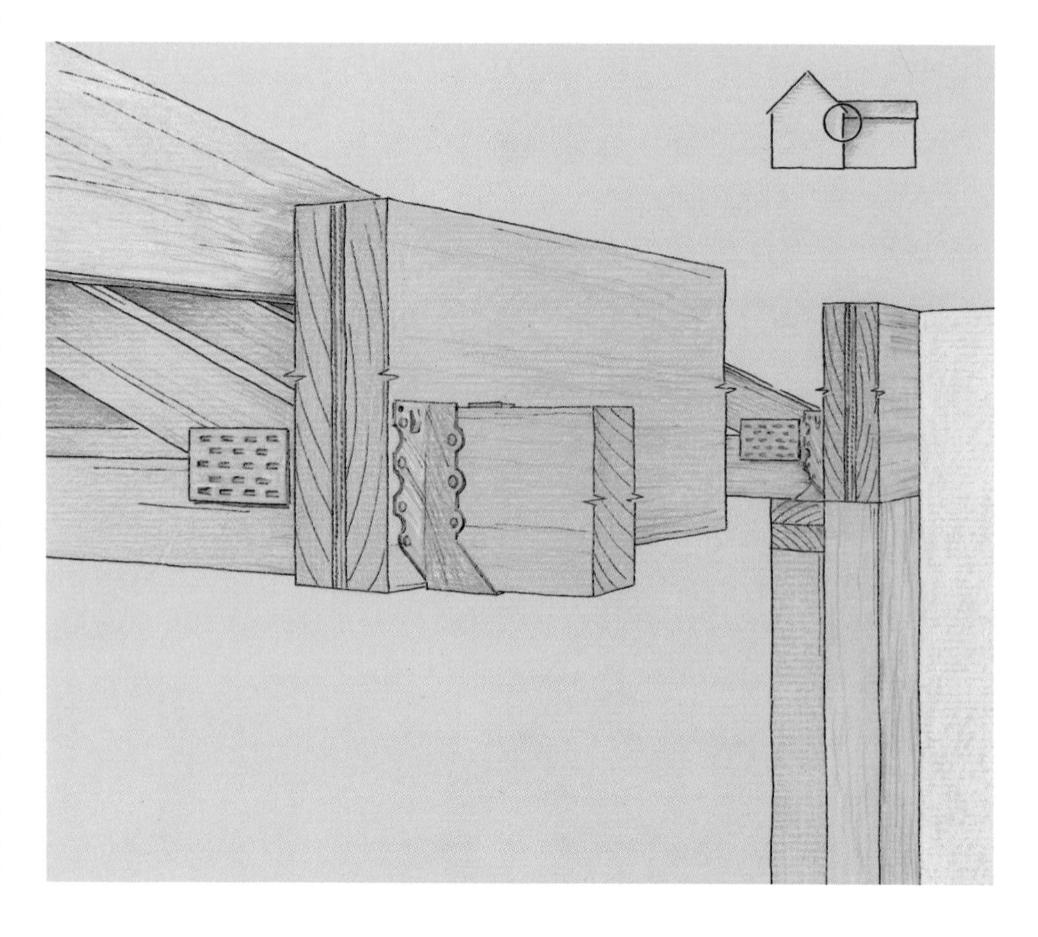

Drawings: Christopher Clapp

tom chord of one truss. Needless to say, the curtain bottom became wrinkled as gravity did its job and the noble truss sagged. The span was too long to double the truss up without adding extra support, so we went up in the attic and set a strongback beam perpendicular to the truss at the center of the span. Our beam also sat on three trusses on each side of the beleaguered curtain-bearer. We were told we couldn't use posts in the dining rooms, so we fastened the trusses to the beam with twisted metal straps, then fastened small cables to the beam and to the top chord of each truss. Structurally, this joined the beam, the top chord and the bottom chord into a single unit. Each cable included a turnbuckle somewhere along its length. After tightening the turnbuckles, we installed additional 2x framing between the beam and several top chords to lock everything in place.

Minimizing drywall damage—The final header method I'll discuss does result in a visible beam, but it saves the wallpaper. We needed to cut an opening in an existing wall. On the side of the wall that had painted drywall, we cut that drywall where the top of the new header would be (drawing 4). On the wallpapered side, we cut the drywall out where the bottom of the new header would be. All the waste drywall was then removed, exposing the studs. We cut the nails holding the studs to the top and bottom plates, then carefully twisted each stud so that the drywall nails pulled out through the back of the wallpapered drywall. After setting posts in place at each end of the opening, we applied glue to one side of a 2x and slid this in place against the back of the wallpapered drywall. This invisibly fastened the wallpapered drywall to the header. We then set the second half of the header in place and *screwed* the two sides together to keep from disturbing the existing drywall. Wrapping the opening with wood trim left the wallpapered side as finished as ever.

There are other methods of hidden-beam installation, but I won't know what they are until the next owner gives us the next set of requirements and the wall is opened up. □

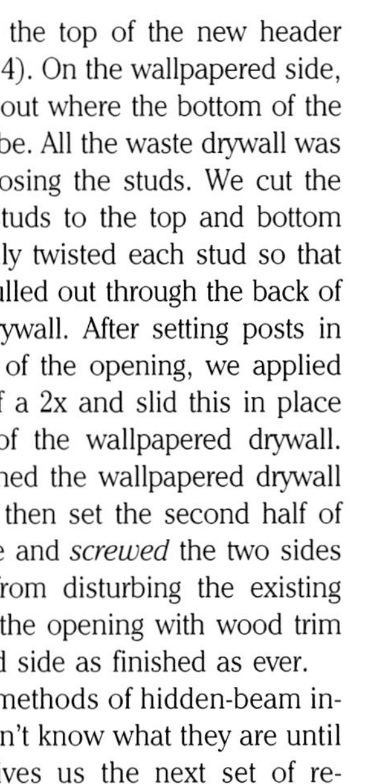

Roger Gwinnup is a builder who recycles old houses near Iowa City, Iowa.

4. Minimizing drywall damage

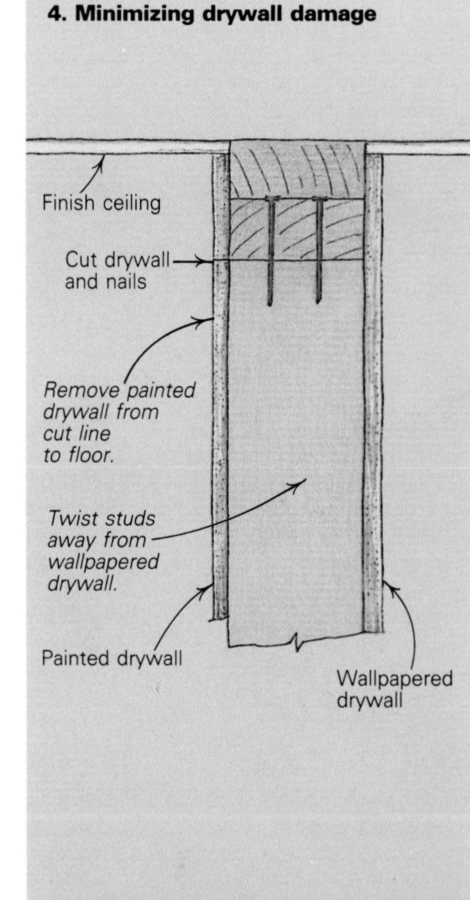

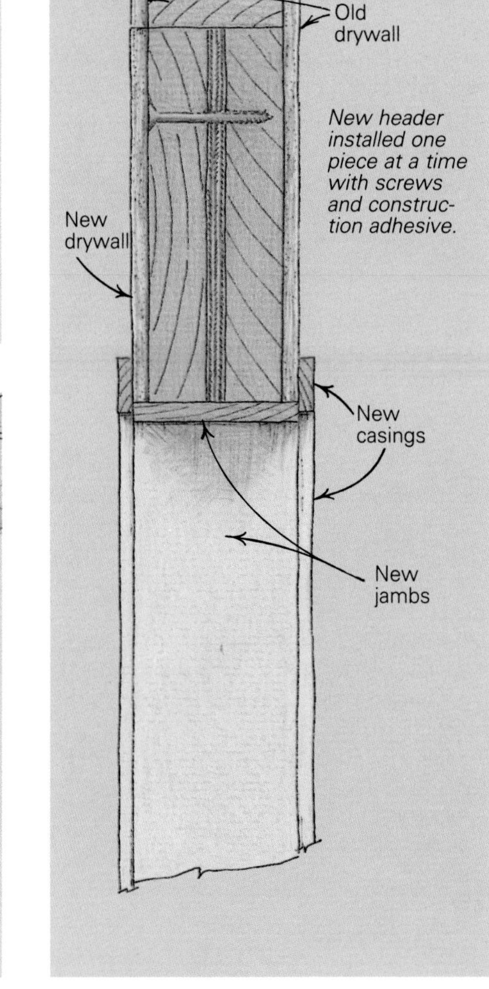

2. Balloon-frame addition

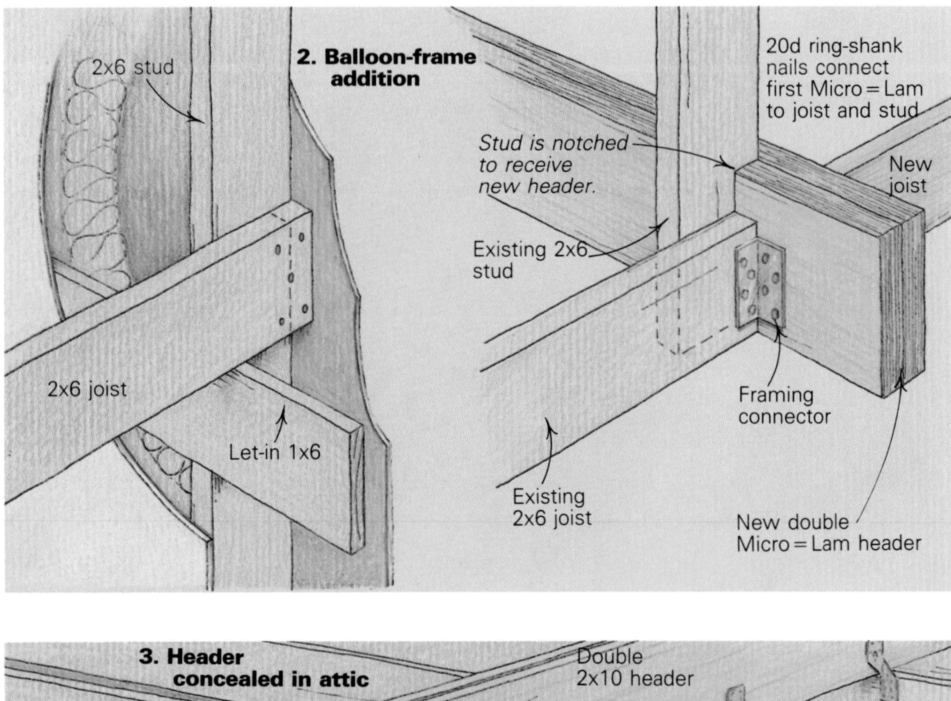

3. Header concealed in attic

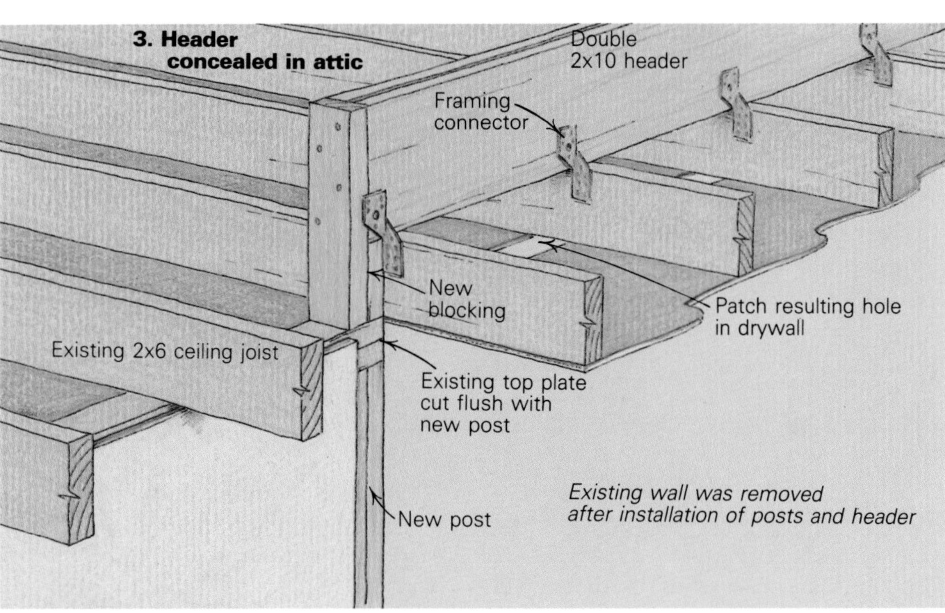

Problem solvers

Concealed header

I needed to remove an 8-ft. bearing wall, but didn't want the new header to break the ceiling surface between rooms (drawing 5). Once the ceiling was shored up, the wall removed, and the joists exposed, I cut slots the thickness of doubled 2x10s, plus a spacer, through each joist. Because I was working alone, I pushed each 2x10 separately up through the gap, into the attic and supported each end on the top plates of opposing walls. Next, I connected each joist to the new header with joist hangers. Then it was a simple matter to patch the surfaces where the wall had been. Like magic, there was no trace of any structural element.

—Steve Orton, a builder who lives in Seattle, Wash.

Heavy long-span headers

Installing a long-span header is an inherently dangerous process. A 12-ft. to 15-ft. flitch-plate header or steel I-beam can weigh several hundred pounds. Here's how I easily and safely install a steel I-beam, using only one helper, some scrap lumber, and a couple of tools.

First I install two pairs of guides made of two 2x's between the ceiling and the floor at each end of the beam (drawing 6). The space between the guides should be about ¼ in. wider than the width of the beam. I then drill a ¾-in. hole through both sets of guides about 3 in. below where the lower edge of the installed beam will be. If the beam is heavy or I haven't had a good breakfast, I'll add some extra holes every foot or so. I use four sill-plate L-bolts or lengths of rebar in the holes for temporary beam supports. These allow me to lift one end of the beam and then the other as I "ratchet" it toward the ceiling. The advantage to this method is that the beam can't tip and I can stop any place I need to. To get the last two or three inches, I use a 4-ton hydraulic bottle jack and a scrap 4x4 post. This allows me to preload the beam enough so the cripple studs don't need to be bashed into place. Sometimes the building groans a little when I remove the supports from underneath the beam. The steel beam is usually straighter than the wall it replaces and the building has to adjust itself to the change.

—Roy K. Jenson, a house remodeler in Edina, Minnesota.

Lateral support for long-span headers

As a building inspector, I'm concerned about lateral movement in long-span headers. Fastening a header in place by toenailing it into the upper plate works for a short span, but I have my doubts about this technique for headers longer than 6 ft. to 8 ft., a span that is too great without some consideration for "racking." While this should be a design concern in every structure (to counteract wind loads, for example), it is of special concern in seismically active areas.

To solve this problem, I usually ask the builder to install ½-in. plywood on both sides of the wall at either end of the header (drawing 7). This is called a shear diaphragm. The plywood should extend uncut 12 in. to 16 in. onto the header from each end to form an inverted "L." Shear walls constructed of plywood must be a minimum of 5/16 in. thick for studs 16 in. o. c. Six-penny common nails, 6 in. o. c., are the smallest permissible size to be used in a shear panel. Nails at panel edges should be no less than ⅜ in. from the edge and no greater than 4 inches apart. Where a shear wall will not work, metal framing straps can be used in innovative ways to help provide lateral stiffening.

—Lee Braun, a building inspector in Belvedere, California.

Headers and point loads

One important effect to consider when installing long-span headers is that of newly created point loads. The load which was once evenly distributed along the bearing wall is now concentrated at two points. In many cases the original foundation may not be capable of supporting these concentrated loads without the risk of cracking or differential settlement. The existing footing and soil conditions should be investigated by an architect or engineer to determine the foundation's ability to support these new point loads. If there's a problem, the situation should be remedied *before* the point loads are created. I suggest that the structural situation be assessed by an architect or engineer when dealing with any span of 7 ft. or longer.

—Martin Hammer, an architect in Oakland, California.

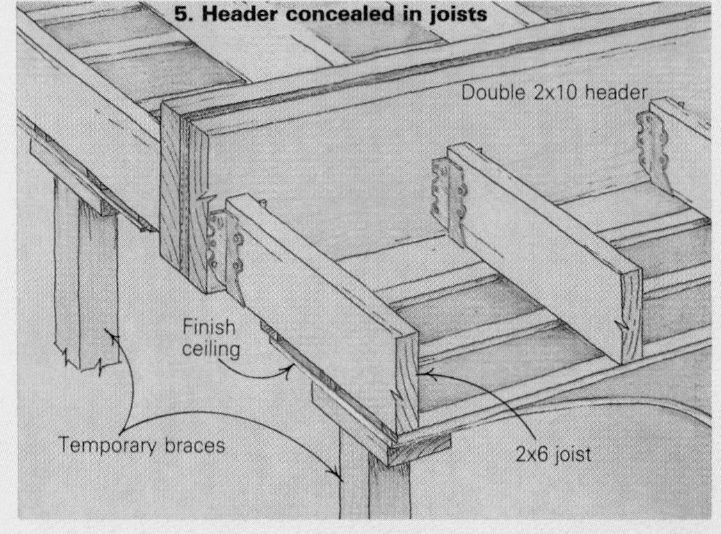

5. Header concealed in joists

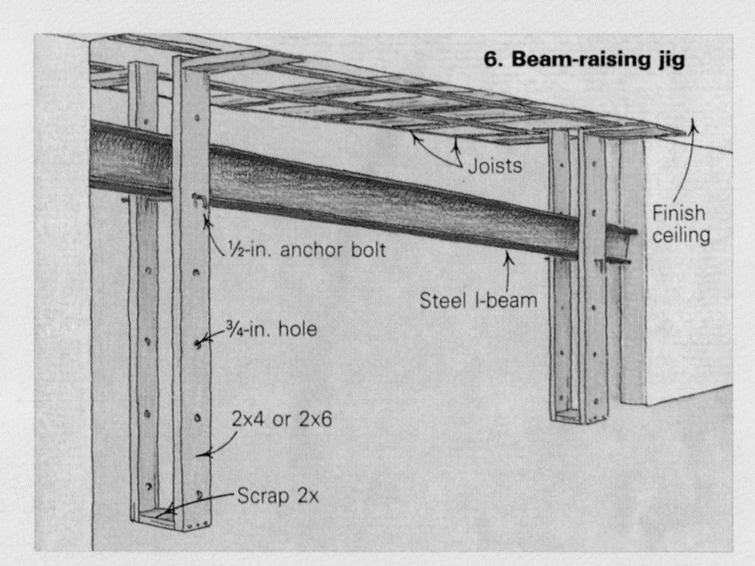

6. Beam-raising jig

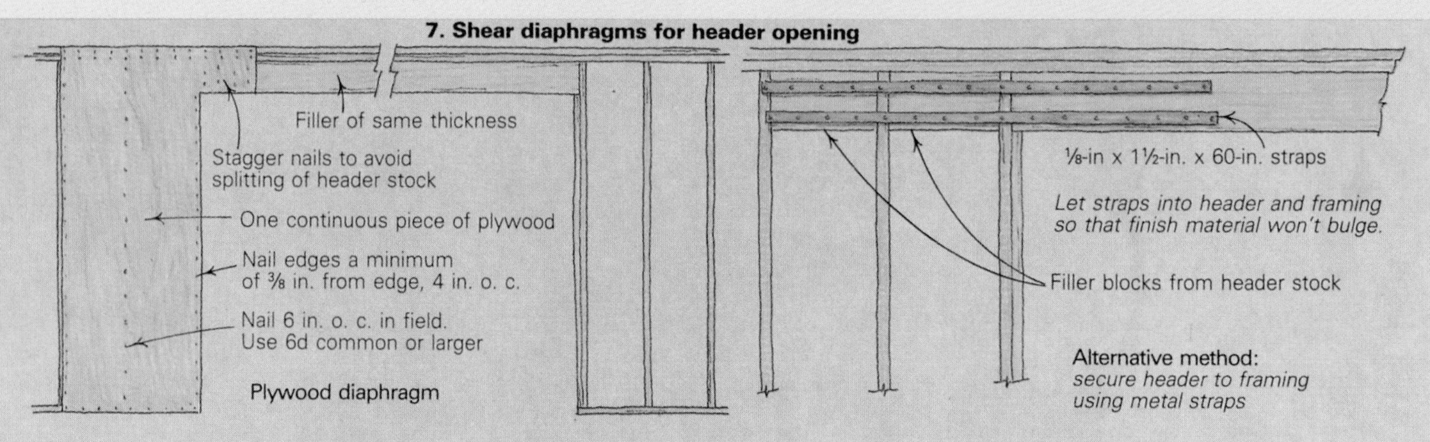

7. Shear diaphragms for header opening

Installing a Long-Span Header

How to open up an existing bearing wall

by Matt Holmstrom

One of the most dramatic ways to alter a living space is to tear out a wall or cut a large opening in it. "Open up the floor plan" and "let more light in" are catch phrases every remodeler hears frequently. But many homeowners and novice remodelers find this an intimidating prospect. The work is extremely messy and seems chaotic. And two words—"bearing wall"—keep many people from tackling it. Actually, this is a straightforward job that demands more common sense than technical skill. Whatever the circumstances, the same basic procedure is followed. Advance planning, careful observation and a step-by-step approach will eliminate that "I'm in over my head" panic.

This view is from the newly framed addition toward what had been the exterior wall of the house. My clients wanted the wall opened up in order to turn the rooms into a single space. Note the various explorations that offer clues as to what lies beneath the sheathing.

Know thine enemy—Simply put, a bearing wall is a wall that bears some of the weight of the structure above it. A wall that is not load bearing supports only its own weight and that of the finish materials on it. To remove a non-bearing wall, just demolish it. Removal of a bearing wall, however, will require some temporary structure (called shoring) to support loads while the work is going on, and will require the installation of a permanent load-bearing member (usually a header or beam supported by posts or studs) to take the place of the removed portion of wall. If you expect to have a framed opening by afternoon where there was a wall that morning, much of the work will have to be done before the reciprocating saw comes out of its case.

First, determine if the wall to be altered is in fact a bearing wall. This is usually fairly easy. Generally, a bearing wall is perpendicular to the joists and/or rafters above it. The weight supported by the joists is sitting on the wall. Another clue is to look beneath the wall (in the basement or the crawl space) for indications that the wall is transferring loads to the ground. You might find another framed wall, a beam set on piers or a foundation wall. If the suspect wall is on a second floor, you'll have to figure out what supports it and then look to see what's beneath *that* wall.

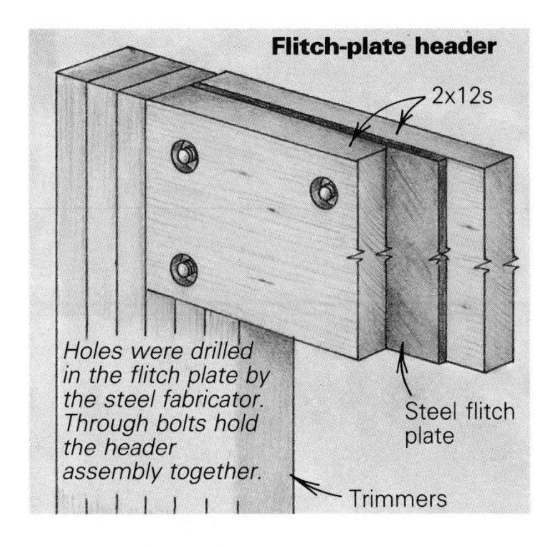

Sizing the header—The next step, assuming that you are indeed dealing with a bearing wall, is to figure out approximately how much wall you'll need to remove and then size the header. The carpenter's rule of thumb used to be this: for spans 4 ft. or less, the header was made of doubled 2x4s; for up to 6 ft. of span, doubled 2x6s; and so on up to 12 ft. of span and 2x12s. Nowdays, however, the building inspectors in my area always want to see at least 2x10s in a bearing-wall header, so I use these for anything up to 10 ft. Given that the structural integrity of the house depends on correctly sizing this header, I'd recommend that you check with your local building department if you're at all in doubt. In any case, headers less than 12 ft. in length don't call for anything fancy, and you can get the materials at the local lumberyard.

When you're dealing with header spans greater than 12 ft., or if the header will be supporting unusual loads (a large bathtub, perhaps, or a slate roof), you'll need to plan extra carefully. The options for these headers include a steel flitch plate bolted between faces of 2x stock, a steel I-beam, a glue-laminated beam, or perhaps a truss. Cost, availability, delivery time, weight and the available headroom are all factors to consider. A structural engineer or an architect can help you with this decision. In fact, your local building official may require that you consult an engineer or architect before proceeding. With any of these manufactured headers, exact span measurements are crucial—you don't want to trim ½-in. steel plate at the job site if you don't have to. Also, you'll have to allow additional lead time to obtain the header.

Checking the wall—The last part of your preliminary work is to check the wall for mechanical systems and to coordinate with the proper subs if necessary. Wiring, plumbing, and heat ducts may be in the wall and, with a few exceptions, will have to be eliminated or rerouted before the header can be installed. Once you start removing structural members, you can't dilly-dally around waiting for the electrician to show up. If you have the luxury of working in an unoccupied house, you can strip off the wall surfaces now and find out what you're dealing with. Otherwise, some detective work will be necessary.

Water pipes and heating ducts are pretty easy to track. If there is much plumbing in the wall, you've picked the wrong wall to tear out—it gets to be quite a job. As for wiring, wall outlets in the work zone are not always a problem. If wires come up through the sole plate, outlets can be left alone until the struc-

Drawing: Michael Mandarano

tural work is completed. But any wire that comes down through the top plate will have to go right away; that usually means switches will have to be relocated.

A case study—Once all the preliminary work is done, you're ready to proceed with wall removal. By now you know approximately what size opening you are going to cut in the bearing wall; the header materials are on hand; and you have dealt with, or are prepared to deal with, any mechanicals in the wall. Things go quickly now. In one or two working days, depending on the complexity of the job, you'll have a new opening ready to finish. I recently opened up a bearing wall and replaced a good portion of it with a header. Here's how it worked out.

With the header in place, trimmers were quickly positioned and nailed to the king studs. Then the header was toenailed to the framing.

We were called in to build an addition to a brick-veneer ranch house. To make the month-long project easier on our clients, we removed a portion of the brick veneer, built the addition, then removed the load-bearing wall between old and new (photo previous page). That way, the exterior wall was never opened to the outdoors. Because the house had a hip roof, all exterior walls were bearing walls. In addition to supporting the old roof rafters and ceiling joists, the wall we removed would have to support the new ceiling joists of the addition.

In order to open the dining room to the new addition, we would have to replace most of the bearing wall with a 15-ft. header and appropriate support framing. One end of the header would rest on new studs added to what remained of the original wall; the other end would sit in a framed pocket formed by the junction of the old wall and the addition wall. One framing concession to our later tie-in had been building the floor of the addition slightly off level to match the existing floor; we didn't want to draw attention to the juncture of old and new.

For the header I opted for a ½-in. steel flitch plate bolted between a pair 2x12s (drawing previous page). The plate cost about $150, and our local steel fabricator had it ready on just a few days' notice. Because the finished opening would be approximately 15 ft. wide, I ordered the plate 15-ft. 9 in. long. That would allow 4½ in. of support under each end, which is what our building inspector asked for. The 11-in. width of the plate would allow us to fit it completely within the depth of the 2x12s.

Temporary 2x shoring in the foreground of the photo below was placed to support the ceiling loads before the structural portions of the bearing wall just behind it could be removed. Lifting the flitch-plate header into place called for plenty of manpower and some well-choreographed moves. The left end of the header fits into a framed pocket in an existing wall, while the right end will be supported on new 2x framing.

The wall contained a few switches, an outlet, an exterior light fixture and two hot-air supply ducts in the portion we planned to remove. The electrician eliminated all these wiring circuits when he roughed in the wiring for the addition. If you plan to leave any wiring in a wall during demolition, however, find the panel box and shut off all circuits to the area before beginning demolition. As for the hot-air ducts, we decided to disassemble them once the wall was stripped; the HVAC guy would later reroute them to supply the new addition.

Stripping the wall—It's incredible how much mess and debris even a small demolition job creates. I try to isolate the work area from the rest of the house and minimize the mess as best I can. Masking tape held 6-mil plastic sheeting over every opening that led to the rest of the house. If there are appliances or large pieces of furniture that can't be moved from the work area, I cover them with plastic sheets or drop cloths. I tape red rosin paper over nearby finished floor surfaces because plastic sheets are just too slippery—they're not tough enough, either. Besides, the paper is cheap and fairly tough, and the 3-ft. wide rolls are easy to handle. To absorb direct hits from dropped tools and falling debris, I lay a sheet of plywood over any finished floor adjoining the wall. Now's the time to remove all trim, doors, hardware or anything else that you want to save. Make sure you store them in another location, too. Once you start tearing into the wall, a certain inertia of demolition takes over and anything can just disappear in the debris.

This house had fiberboard sheathing on the exterior side of the wall to be removed and plaster over rock lath on the interior. I drew a rough layout of the opening directly on the plaster, allowing more than enough length for the header and a few studs ganged on each end; this determined where to cut the plaster. After removing trim and the existing window and door, we tackled the work. We stripped the plaster right up to the ceiling along the entire length of the header, so we had a good view of the doubled top plate our header would be supporting. After cutting plaster, we always clean up the mess to avoid grinding plaster dust into the finish flooring (oak in this case); protective paper can't always contend with the fine, gritty powder left by this kind of demolition (that goes for drywall demolition, too). A shop vac is almost a necessity here.

Building the header—With the wall framing exposed (but not yet cut), the new header, king studs and trimmer studs can all be laid out, and the header can be built. A header of this length (15-ft. 9 in.) and weight needs three trimmer studs on each end for support. Normally, if the exact position of the new wall opening is not critical, I try to use an existing stud as one king stud for the new header, and I begin my final layout off this.

From *Fine Homebuilding* magazine (August 1989) 55:75-77

Here, the kitchen-window stud was my starting point. This was 6⅞ in. back from the drywall face of our perpendicular addition wall. I sistered a new stud against this old one, shimming between them to get it plumb. This would be the king stud for the new header. The remaining 5⅜ in. between this stud and the intersecting wall face could be filled nicely with three trimmer studs (4½ in.), a ⅜-in. filler of plywood or drywall and ½-in. drywall. The result would be a nice outside corner where the walls intersect. Then I toenailed the opposing king stud into place 189 in. away from the first king stud. I left the bottom plate in place for now.

The toughest part of this project turned out to be moving the steel flitch plate. It weighed 295 lb., and it took two men one-half hour to slide the plate off the truck racks and maneuver it onto sawhorses without damaging fingers or backs. Assembling the header was comparatively easy. I had hand-picked two straight 16-ft. 2x12s at the lumberyard. You want as little crown as possible in any header, but here even ⅛ in. of crown would have been difficult to deal with when it came to fitting the header into position. Any problems in fitting such a long, heavy header could mean more than wasted time—it could result in personal injury.

The flitch plate had been predrilled by the steel fabricator, so assembly was simply a matter of marking the hole alignment on the 2x stock, drilling and countersinking the holes, and securing the whole affair with ⅜-in. by 3½-in. bolts, nuts and washers. We lugged the completed header into the addition and set it at the base of the stripped wall. The only way two men could move this monster was by sliding it along "leapfrogged" sawhorses up to the door, and then sliding it along the addition subfloor. Before we could begin to remove the last of the old wall, however, we had to shore up the ceiling to support ceiling and roof loads.

Setting the shoring—There are two ways I know of to build shoring, and I used both on this job. One method calls for building a 2x6 stud wall to support loads; we did this in the addition (top photo, facing page). Line up the studs under every second joist you have to support (every 32 in. o. c. in this case). The second shoring method calls for a beam (two sistered 2x6s) and two or three posts to support it against the ceiling loads; we did this in the dining room. The first method takes longer to build, but is probably more stable than the second. It spreads the load better, too. I use the second method under an uneven plaster ceiling—it minimizes ceiling damage, and I don't have to spend time locating ceiling joists in the plaster. To be effective, the shoring must be snug against the ceiling, but not so snug that it causes damage.

I usually use 2x6s for shoring walls. They're a little more rigid than 2x4s, and less likely to split when you bang them into place. The shoring should be set about 2 ft. from the bearing wall so that there's plenty of room to maneuver a stepladder between shoring and wall.

After the shoring was in, we completed the demolition by cutting away the rest of the wall studs. The easiest way is to cut each stud in half with a reciprocating saw and remove it, then cut the nails protruding from the top and bottom plates. Use bi-metal blades in your reciprocating saw (I buy ones labeled: "For nail-embedded wood"); they cost more but last much longer for this kind of work. If you bend a blade (and you will), simply straighten it with pliers and get back to work.

Installing the header—You'll often be working in a small, cluttered area, so choreograph the installation: who will be where, which end of the header will go in first. I used a disc sander to bevel slightly one end of the header along its width. This made it considerably easier to slide the header between the king studs. After recruiting help in the form of two carpenter friends who were working nearby, I leaned the precut trimmer studs against the wall near each end of the opening. The rest happened fast: the four of us (with a bit of help from a friend) lifted the header up and slipped one end in first, while a man on each end knocked the first trimmers in place and quickly tacked them to the king stud (bottom photo facing page). Then the header was driven the rest of the way in with a 3-lb. hammer and a scrap block, and an additional two trimmers were nailed off on each end, with two more studs behind the king stud. After toenailing the header into the top plate, we cut out the bottom plate flush with each end trimmer and pulled up for a rest. Our opening was framed and would be ready to finish once the shoring came down. □

Matt Holmstrom is a remodeling contractor who prefers to work on older homes. Photos by Bill Hoy.

The effect of removing a wall is dramatic. In this case, the original front door was directly in front of this basement door.

Framing a Bay Window With Irregular Hips

How one carpenter calculates the tough cuts

by Don Dunkley

My crew and I frame houses in central California, near Sacramento, where designers compete with one another to see who can create the most complicated roofs. To stay in business, the local carpenters have to be adept at framing every type of roof—hip, gable, octagon, cone—sometimes all in the same building.

One modest and enduring feature that turns up in many of these homes is the bay window popout. The kind we build most frequently, and the subject of this article, consists of two 45° corners and a projection, or offset, of 2 ft. (floor plan, p. 86). It is 10 ft. wide at the wall line and 6 ft. wide at the front of the offset. The plate height of the bay and of the adjoining room are 8 ft. 1 in., and the roof pitch is 8-in-12.

A hip roof commonly tops this kind of a bay. But unlike many of the hip-roof bays that get built locally, we frame ours with two irregular hips (photo above right). More often than not I run across roof plans that leave out a second irregular hip. Without it, the plane of the roof has to be warped to intersect the valley (photo above left). Once you become aware of this refinement, chances are you'll spot many an example of incorrectly framed bays on a casual drive down a residential street. Adding the second irregular hip allows the roof planes to meet at crisp angles.

Building a roof with one pair of irregular hips is a challenge—add another pair and it's a task for a journeyman carpenter. When I first started out as a framing carpenter, I spent a lot of time laying out the rafter locations on the subfloor, then transferred them by way of plumb bobs and stringlines to temporary staging, where another session with stringlines and tape measures would follow as I puzzled out seat cuts and cheek cuts. No more. I've incorporated two tools into my roof-cutting procedures that do away with all the plodding.

The first is a Construction Master Dimensional Calculator (Calculated Industries, Inc., 22720 Savi Ranch Pkwy., Yorba Linda, Calif. 92686). This calculator works in decimal numbers and in feet and inches. It also has pitch, rise, run, diagonal and hip/valley functions, which eliminate some of the key strokes required to apply standard calculators to carpentry work. For instance, instead of using a square-root formula to find a diagonal, I enter 11 ft. as rise, 14 ft. as run, punch the diagonal button, and the calculator will read 17.8044. When I punch the convert-to-feet-and-inches button, it tells me 17 ft. 9⅝ in.

The second tool is a book called *Roof Framing*, by Marshall Gross (Craftsman Book Company, P. O. Box 6500, Carlsbad, Calif., 92008). Gross uses a technique to lay out roofs that he calls the "height above plate" method (HAP). Simply stated, the HAP system allows me to set the ridges first at their actual height. Then I bring the rafters to meet them. I've found this system to be unbeatable for assembling complex roofs. But before we dive into HAP and bay window/roof theory and calcs, let's look at layout and walls.

Leaving the second irregular hip out of a bay-window roof causes an awkward warp in the sheathing, which shows up in the shingling and the valley flashing (photo above left). The bay in the photo at the right was framed with both hips, and the valley runs straight and true.

Patterns and plates—We build bays like the one shown in the photograph (top photo, facing page) on either slab or wood-framed floors, and sometimes on cantilevered joists. In each case, laying out the bay begins after the subfloor is in place and I've snapped a chalkline marking the inside edge of the wall plates around the house.

To save time and ensure accuracy when I mark the position of the bay, I use a plywood pattern of a 2-ft. by 2-ft., 45° corner (floor plan, p. 80). By placing the pattern along the wall line at the beginning of the bay, I can quickly lay out a perfect 45° wall. The pattern has layout marks on both sides, so I can flip it to lay out the opposite corner.

I usually make square cuts on the ends of the diagonal wall plates. They abut outer wall plates that have two 45° cuts on their ends (floor plan detail, p. 86). I do this for two rea-

From *Fine Homebuilding* magazine (December 1989) 57:78-83

sons: the angled stud on the outer wall plate gives me a good nailing surface to anchor the walls together, and it gives me a little more room to squeeze the window into the diagonal wall. Designers inevitably want lots of windows in these walls. That means I need from 26½ in. to 27½ in. for my header, depending on the width of the windows. As you can see from the photo below, this can get snug. To make the windows fit, I sometimes have to use 1x4 king studs instead of 2x4s next to the trimmers that carry the window headers.

After the walls have been framed and plumbed, it's time for the roof. If you're familiar with roof theory, I'll go straight to calculating the rafters for the bay. If you'd like to brush up on roof basics, please refer to the sidebar, "Regular and irregular hips," on p. 89.

Locating the ridge—To find the ridge height using the HAP method, I add the distance the rafter sits above the plate at the seat cut to the theoretical rise, minus the reduction caused by the thickness of the ridge (elevation view, p. 87). For example, our seat cut (the horizontal portion of a bird's mouth) is 3½ in. on the level, giving a 4¼-in. rise above the plate for a 2x6 rafter. The run of the common rafter is 5 ft., and the rise is 40 in. (8-in. pitch by 5-ft. run). This gives us a theoretical rise of 44¼ in. If there were no ridge, the peak of the rafters would be this height, but the ridge comes between them and must be accounted for. This applies to both common rafters intersecting the ridge at a right angle or, as in this case, a common rafter in line with the ridge. I find the reduction the "new-fashioned" way, courtesy of Construction Master (detail 1, p. 87). Our ridge is 1½ in. thick. Using my Dimensional Calculator, I enter half the thickness of the ridge (¾ in.) as the run. Next I enter 8 in. as the pitch, and punch the rise button. My answer is ½ in. I subtract that from 44¼ in. to get the actual height of the ridge above the plate: 43¾ in.

To begin erecting the bay's roof, I set its ridge on temporary legs so that it sits precisely 43¾ in. above the plate. If I later cut all my rafters accurately, and my walls have been properly plumbed, lined and braced, all the parts will converge to lock the assembly together.

Because the offset of the bay is 2 ft., the ridge is approximately 2 ft. long. Actually, it's a little longer in order to compensate for the shortening allowance of the common rafter. More on this in a minute.

First, the valleys—Our floor plans show a 2-ft. offset and a 6-ft. long front to the bay window. The interior opening of the bay is 10 ft. wide. The roof plan shows two valley rafters, one common rafter and two sets of irregular hip rafters (roof plan, next page). The roof overhang is 2 ft.

At this stage of the roof framing I'm not concerned about the 45° wall of the bay. Instead, I'm thinking about the 10-ft. wide opening to the bay. Dividing it in half gives me a pair of 5-ft. squares in plan. The bay's ridge and common rafter form a line between the squares, and its regular valleys are the diagonals.

An 8-in-17 valley (see sidebar) on a 5-ft. run calculates to be 7 ft. 9¾ in. long. To get my 2-ft. overhang, I have to add 3 ft. 1½ in. from the seat cut to the tail cut. The vertical edge (heel cut or plumb cut) of the bird's mouth aligns with the exterior face of the wall framing.

The skinny little lines we see when we look at roof framing plans represent the center lines of rafters and beams. To transfer the ideal of a line with no width into a rafter that is typically 1½ in. thick, we have to take the shortening allowance (SA) into consideration. For a regular hip or valley, the SA is equal to half the thickness of the common rafter, cut on a 45° angle (detail 2, p. 87). That works out to 1$^{1}/_{16}$ in. for 2x framing lumber. Remember this is a level measurement, and has to be adjusted for the pitch of the roof. For our 8-in-17 pitch, the valley rafter has to be shortened by 1$^{3}/_{16}$ in. To find the adjusted SA with the Construction Master, enter the pitch as the decimal .47 (8-in. rise divided by 17-in. run). Then enter 1$^{1}/_{16}$ in. as the run, and punch the diagonal button to get the adjusted SA of 1$^{11}/_{64}$ in. Round it off to 1$^{3}/_{16}$ in.

I'm in the habit of cutting double cheek cuts on valley rafters. Often there are other rafters intersecting the same ridge, and the double cheek cut gives me a little extra room for adjustment. In addition, to save time at the cutting table I put double cheek plumb cuts on all my valley and hip stock at the same time, and then decide later which ones end up as hips or valleys. The photo on p. 88 shows how the valleys intersect the ridge.

Common rafter—In order for the common rafter to be at the same height as the valley rafter, it also must be calculated on the 10-ft. span, or 5-ft. run. An 8-in-12 common on a 5-ft. run calculates to be 6 ft. ⅛ in. long. Measured on the level, its SA is half the

The author nails down irregular hip rafter A on a cantilevered bay (photo below). Note the 1x4 king studs in the diagonal walls. They allow more room for the window. Opposing walls are joined by a tie beam across the top of the opening to the bay (photo above). The upper half of the top plate has been let into the beam for several feet.

Drawings: Michael Mandarano

thickness of the ridge—in this case ¾ in. (detail 1, facing page).

Our next move is to calculate the lengths of the valley jacks. Since our run from the common rafter to the valley rafter is 2 ft., our valley jack will be cut on that run. The valley jack will have two SAs—one for half the thickness of the ridge and one for half the thickness of the valley measured across its top at a 45° angle.

Irregular hip A—Looking at the roof plan (bottom drawing) we see that the distance from the common rafter to irregular hip A is 3 ft. The run of the common rafter to the ridge is 5 ft. To find the run of hip A, I enter 3-ft. rise and 5-ft. run. When I punch the diagonal button, it reads 5 ft. 10 in. But that's measured on the level. To figure the length of hip A, we need its actual rise. By feeding our 40-in. rise and 5-ft. 10-in. run into the calculator, I get the full unadjusted length of the rafter from seat cut to ridge junction: 6 ft. 8⅝ in.

While we're working with these numbers, let's figure out the plumb cut for this hip by dividing our 40-in. rise by our 5-ft. 10-in. run to get tangent .571428. The trig tables say that's almost 30°. While the tangent number is still in the calculator's display screen, I punch the convert-to-inches button, which now reads 6⅞ in. That means the pitch (and the plumb cut) of this hip rafter is 6⅞-in-12.

Because we're dealing with an irregular hip here, we need to know the angles formed by its intersection with the plate and the common rafter. Without them we can't calculate the cheek cuts on the hip rafter at the ridge or the cheek cuts on the hip jack rafters. Using the tangent method and the trig tables, I find that the angle made by the hip

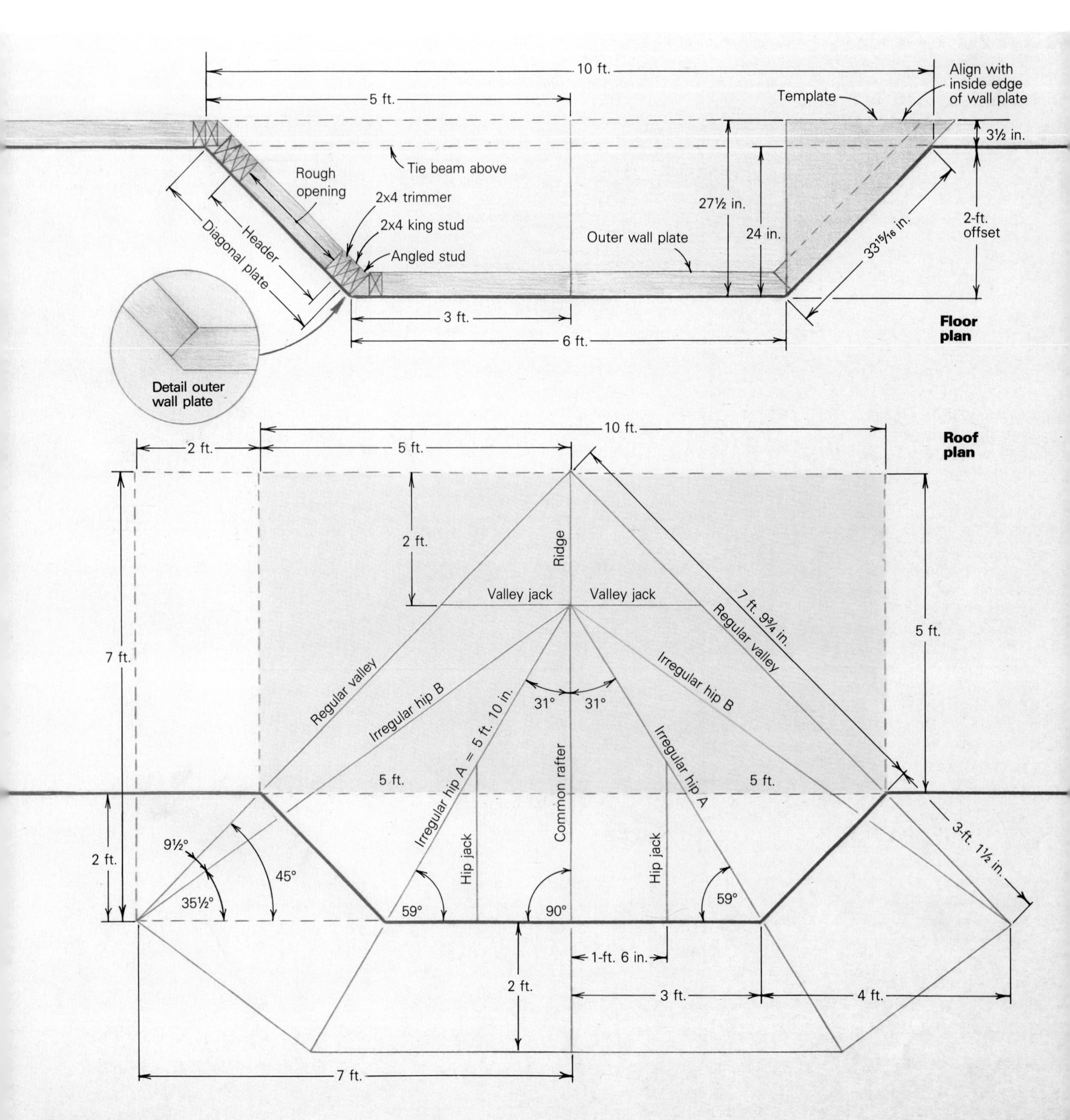

and the plate is 59°; therefore the complementary angle is 31°.

To be an irregular hip is to be off center at the intersection of all the other rafters (plan of rafter intersection, next page). Here's a way to calculate the SA and cheek-cut angles for this asymmetrical junction. In the triangle ABC, the rise of BC is half the thickness of the ridge, or ¾ in. The angle A is 31°, derived from the plan view of our roof. Thirty-one degrees is the same as a 7³⁄₁₆-in. roof pitch. Calculator in hand, I enter 7³⁄₁₆ in. as pitch and ¾ in. as rise, punch the diagonal button, and it reads 1⁷⁄₁₆ in. for the hypotenuse AB in triangle ABC. This is the SA, measured on the level. The adjusted SA for this 6⅞- and 12-pitch is 1⅝ in.

Let's take a look at the rafter intersection plan to see how the framing square is used to lay out this irregular-hip plumb cut. First, measure back from the full length of the rafter the adjusted SA, 1⅝ in., to mark point X on the rafter's centerline. By looking at the plan view we see that the hip needs two different cheek cuts. Let's make the longest side first. By laying the framing square on top of our rafter with the tongue set at 7³⁄₁₆ and the body set at 12, draw a line on the edge of the rafter that passes through point X (detail 3, next page). This line (YD) gives us the angle for the cheek cut on the side of the common rafter.

By studying our plan of our rafter intersection, we see that the two cheek cuts intersect off the centerline of the rafter at point F. To find this point, first square a line from the edge of the rafter to X to find point Z. A line perpendicular to YD that intersects point Z gives us the second cheek cut in plan.

Now we've got the cheek-cut angles for a horizontal rafter. Just to make this a chal-

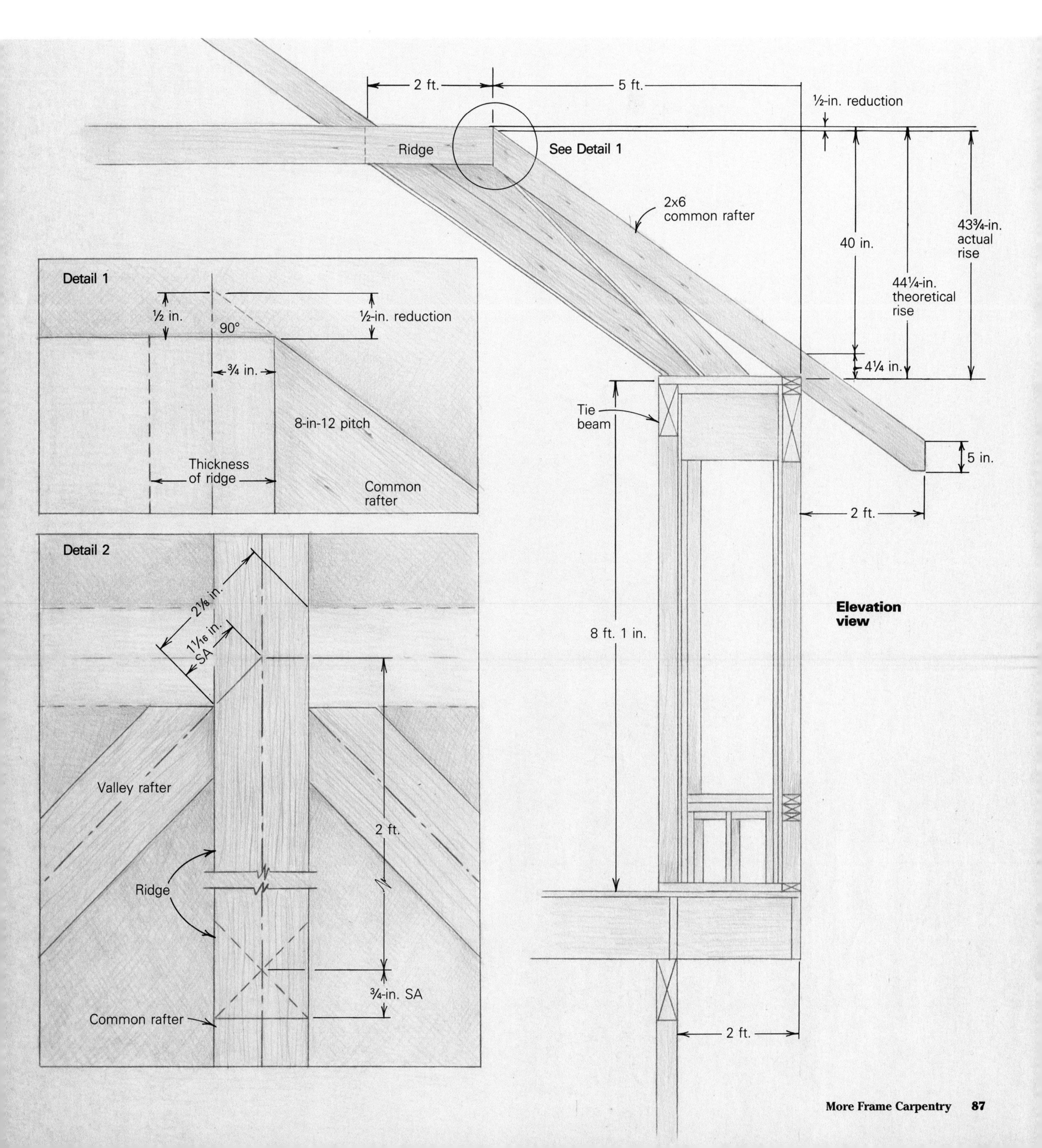

Converging rafters meet at the end of the bay's ridge beam. In the lower left corner of the photo you can see where the valleys intersect the ridge. The "X" marks one end of the 2-ft. ridge.

lenging exercise, the angles change as the rafter's pitch increases—the greater the pitch, the greater the change. You can demonstrate this phenomenon by drawing an equilateral triangle on a slip of paper. Hold the drawing level, with its base toward you. Now slowly rotate the drawing to vertical to change its pitch. You can watch the angle change from an obvious 60° to a right angle and beyond.

When you make compound cuts with a circular saw, such as the cheek cuts on a hip or jack rafter, the saw automatically compensates for the pitch of the rafter. But if you have to use a handsaw to cut an angle beyond the circular saw's 45° capability, you have to compensate for the angle change in your layout. The angle we need to cut here is 59°. Here's how to lay it out.

Recall that our hip plumb cut is 6⅞ and 12. Scribe a line at this pitch on the side of the rafter, beginning at point D (detail 4 below). Now mark from this line the distance EY on the side of the rafter, and scribe another line at 6⅞ pitch. Square this line across the top of the rafter to find point Y1. Connect point F and point Y1 to find the adjusted cheek cut. If you want to cut the other angle with a handsaw, repeat the process to find the adjusted cut for the other cheek. In plan, they look like detail 5 below. Now you're ready to make the double-cheeked plumb cut for irregular rafter A, and to take a break.

In practice, when the rough framing is going to be covered up by a ceiling, I generally trim this cut to fit. A perfect cut isn't necessary for structural integrity, and as always, time is of the essence. But a journeyman carpenter should know how to make this cut precisely if the rafters are going to be exposed to view.

I let the tails of these two rafters run wild past the wall. Once I've got the rest of the rafters in place, I use my level to determine the position of my tail cut. In this manner I can make sure that the fascia and gutters end up in the right place.

Irregular hip B—These hips can be calculated mathematically, but to tell the truth I

Plan of rafter intersection

Ridge
¾ in.
D
E
Valley jack
Valley jack
F
Z
90°
59°
X
1⁷⁄₁₆-in. SA
B
C
EY
Y
59°
A
¾ in.
31°
Theoretical rafter length
Common rafter

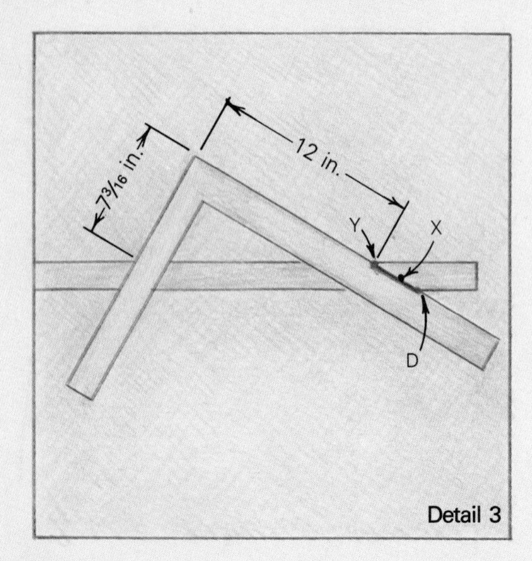

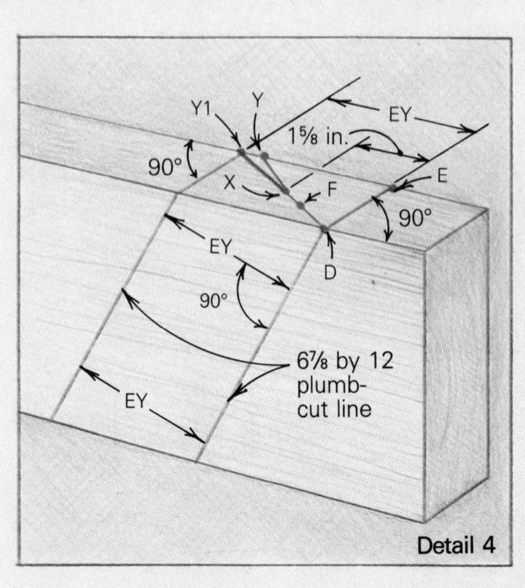

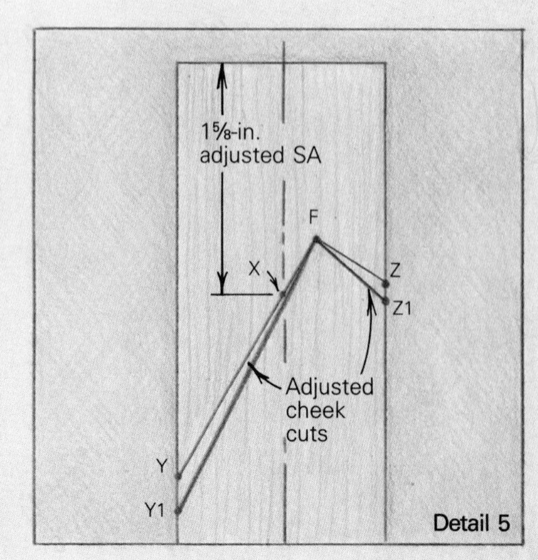

use stringlines to figure them out. I tack a nail in the top of the valley rafter to represent the centerline intersection of the valley and its neighboring hip. Then I run a string to the point at which the rafters are intersecting at the end of the ridge. I'll measure this distance to get the unadjusted length of the rafter, and while I'm at it I measure the distance from the stringline to the wall plate to get the depth of the seat cut.

To lay out the radically tapered tail cut on hip B, I need to know the angle the rafter makes with the front of the bay. Using the tangent method, I find it to be 35½°. From the plan view we see that the 45° valley and 35½° hip come together to form a 9½° angle (45° minus 35½°). The complement of 9½° is 80½°. I use this angle on my Speed Square to mark the tail cut. Once I adjust the cheek cut for the pitch of the rafter, I make the cut with a sharp handsaw. The only rafters left to install are the hip jacks.

Because I have 3 ft. from corner to common rafter, I center the jack at 1 ft. 6 in. You should have enough information now to figure out their rise, run, pitch, length and cheek cuts. □

Don Dunkley is a framing contractor working in California's central valley.

Regular and irregular hips

Understanding complex roofs requires an understanding of the mathematics of a simple roof. Here are some basics.

The *pitch* of a roof is determined by the relationship of vertical *rise* to the horizontal *run*. An 8-in-12 roof means that for every 12 in. of run the rafter will rise 8 in. To represent this relationship visually, a diagonal line connects the two, forming a triangle (detail C below). The diagonal represents the slope of the rafter. Because the rise and run are perpendicular to one another, the three lines form a right triangle. The mathematical formula to find the length of the diagonal of a right triangle is called the Pythagorean theorem: $a^2 + b^2 = c^2$. Another way to figure it: c = the square root of $a^2 + b^2$

Fortunately, the carpenter can reach for a calculator to process these numbers. Another way to bypass tedious rafter calculations is with a rafter book, such as *Full Length Roof Framer* (A. F. Reichers, Box 405, Palo Alto, Calif. 94302). It lists the lengths for common, regular hip and valley rafters for roofs with 48 different pitches.

If you divide the rise by the run, you get the *tangent*. For an 8-in-12 roof, the tangent is .666667. What can you do with this information? By looking at a table of trigonometric functions you'll find that a tangent of .666667 is equal to approximately 33¾°. Therefore an 8-in-12 roof rises at an angle of 33¾°. All roof pitches have a corresponding tangent/degree.

Once we know two of the angles in a triangle, we can subtract their sum from 180° to get the third angle. In our example, 180° - (90° + 33¾) = 56¼°, which is called the *complement* of the 33¾° angle (complementary angles add up to 90°). When laid out with a Speed Square, this 56¼° angle will give the *level* or *horizontal cut* of a rafter whereas the 33¾° is the *plumb cut* (detail A).

In plan, a *regular hip* rafter intersects common rafters at a 45° angle. To understand a hip, look at it from the plan, or top view (drawing below). The common rafters intersect the plates at 90°, while the hip is 45° to the plates. In order for the regular hip rafter to reach the same point as a common rafter, its run must be longer. For every 12 in. a common rafter needs for run, a regular hip rafter, regardless of pitch, needs 16.97 in. (for pitch designations, carpenters round the number off to 17 in.). This relationship holds true for *regular valley* rafters as well. Therefore, a regular hip or valley rafter on an 8-in-12 roof is cut to a 8-in-17 pitch. Divide 8 by 17 to get the tangent: .4705, which gives us the hip *plumb cut*—25½°.

As shown in our plan, a regular-hip roof over a 22-ft. span reveals two 11-ft. squares. The run of the commons will be 11 ft. Find the run of the hips by multiplying the run of the common times the run of a regular hip: or, 11 x 16.97 in., which equals 15 ft. 6¹¹⁄₁₆ in.

So what does this tell us about how to calculate irregular hips? To figure out an *irregular hip* the diagonal length of its run is needed. But because an irregular hip doesn't have a 45° angle in plan, the value of 16.97 can't be used. Back we go to the Pythagorean theorem.

Let's say we have an 8-in-12 roof with commons that run 11 ft., but the commons are 14 ft. from the corner where the hip originates at the plates (drawing below and detail B). By using the Pythagorean theorem, we find that the diagonal in a triangle with 11 ft. and 14 ft. sides is 17 ft. 9⅝ in., which gives us the run of this irregular hip. To find the full length (also referred to as theoretical or unadjusted length) of the hip rafter, use the rise (88 in.) and the run to find the diagonal, which is 19 ft. 3¹⁄₁₆ in.

The hip jacks are another wrinkle—those intersecting an irregular-hip rafter have different angles on their cheek cuts. Using our tangent formula, we find that our irregular hips divide the plan view of our roof into 38° and 52° angles. When the cheeks of opposing jack rafters are cut at these angles and adjusted for the pitch of the rafter as shown in the bottom drawing, p. 86, they'll fit snug against the irregular hip. —*D. D.*

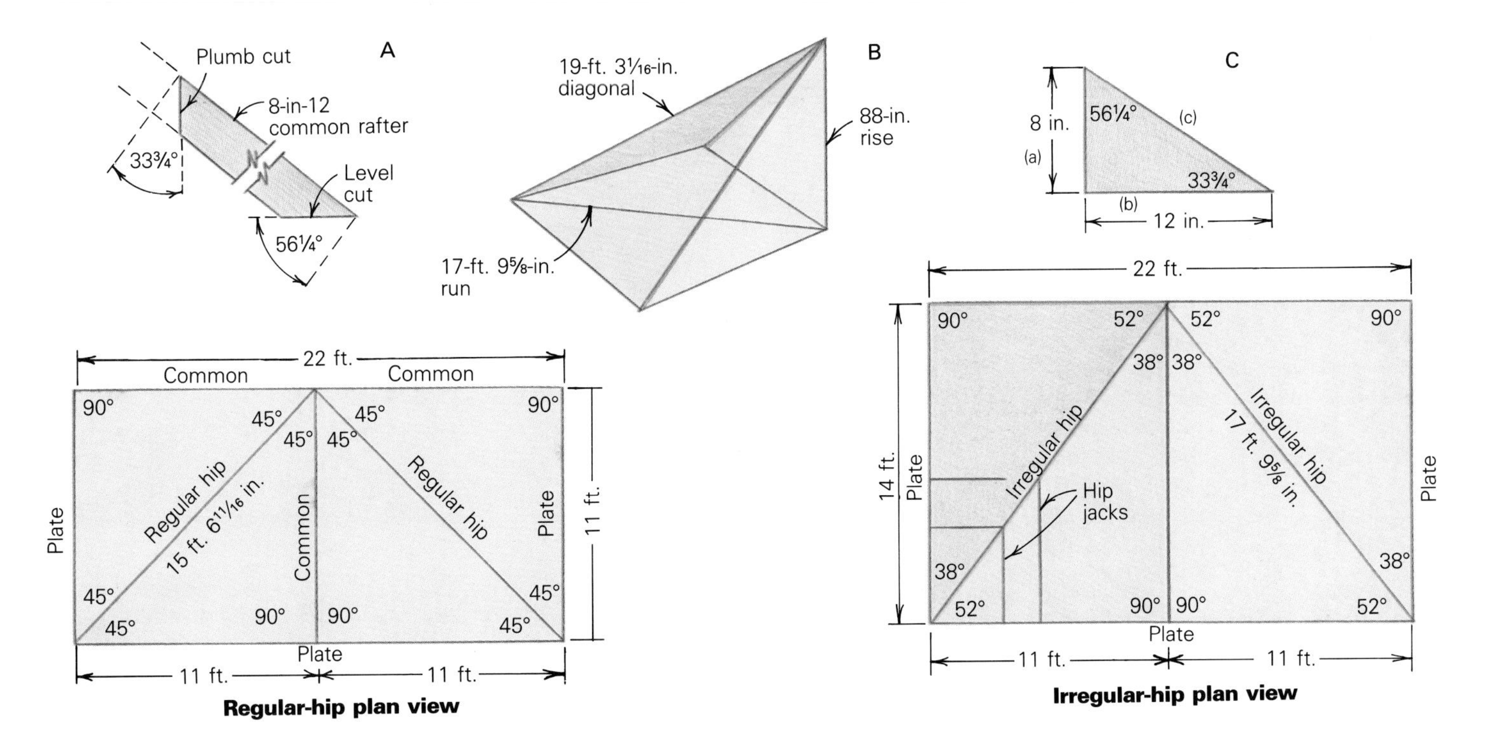

Raising an Eyebrow

Two methods used to frame wave-like dormers

by James Docker

Eyebrow dormers had their American heyday during the late 19th century, when they turned up on the elaborate roofs of Shingle-style Victorian and Richardsonian Romanesque houses. Tucked between the conical towers, spire-like chimneys and abundant gables that distinguish these buildings, the little eyebrows provided a secondary level of detail to the roof and some much needed daylight to upstairs rooms and attics.

The roof cutters of that era could probably lay out an eyebrow dormer during a coffee break, but for a contemporary West Coast carpenter such as myself (well-versed in shear walls, production framing and remodeling techniques), framing an eyebrow dormer presented an out-of-the-ordinary challenge.

The setting for this dormer project was a rambling Tudor house in Atherton, California. The owners were adding a garage and remodeling several portions of the building, including a dilapidated barnlike recreation room next to the swimming pool. The roof of the house was covered with cedar shingles, and at the eaves and gable ends, curved shingles gave the roof a thatched look. Eyebrow dormers, rising by way of gentle curves from the plane of the 8-in-12 roofs, would look right at home on the house (photos right).

My job was to install five of them in the garage roof and a single larger eyebrow dormer in the roof of the recreation room. The garage ceiling would remain unfinished, so I didn't need to worry about providing backing for drywall or plaster. I would, however, have to solve that problem in the recreation room. The garage dormers required a lower level of finish while presenting the same conceptual problems, so I decided to build them first.

Rafter-type eyebrow—By the time I got on the job, contractor Dave Tsukushi had already taken delivery of the windows for the garage roof. They were arched, single-glazed units available off-the-shelf from Pozzi Wood Windows (Bend Millworks Systems, P. O. Box

Undulating courses of cedar shingles wrap over the warped contours of these eyebrow dormers on a Tudor-style house in Atherton, California. The solo dormer (top photo) lets light into the recreation room, while the others illuminate the garage (bottom photo).

Photos this page: Charles Miller

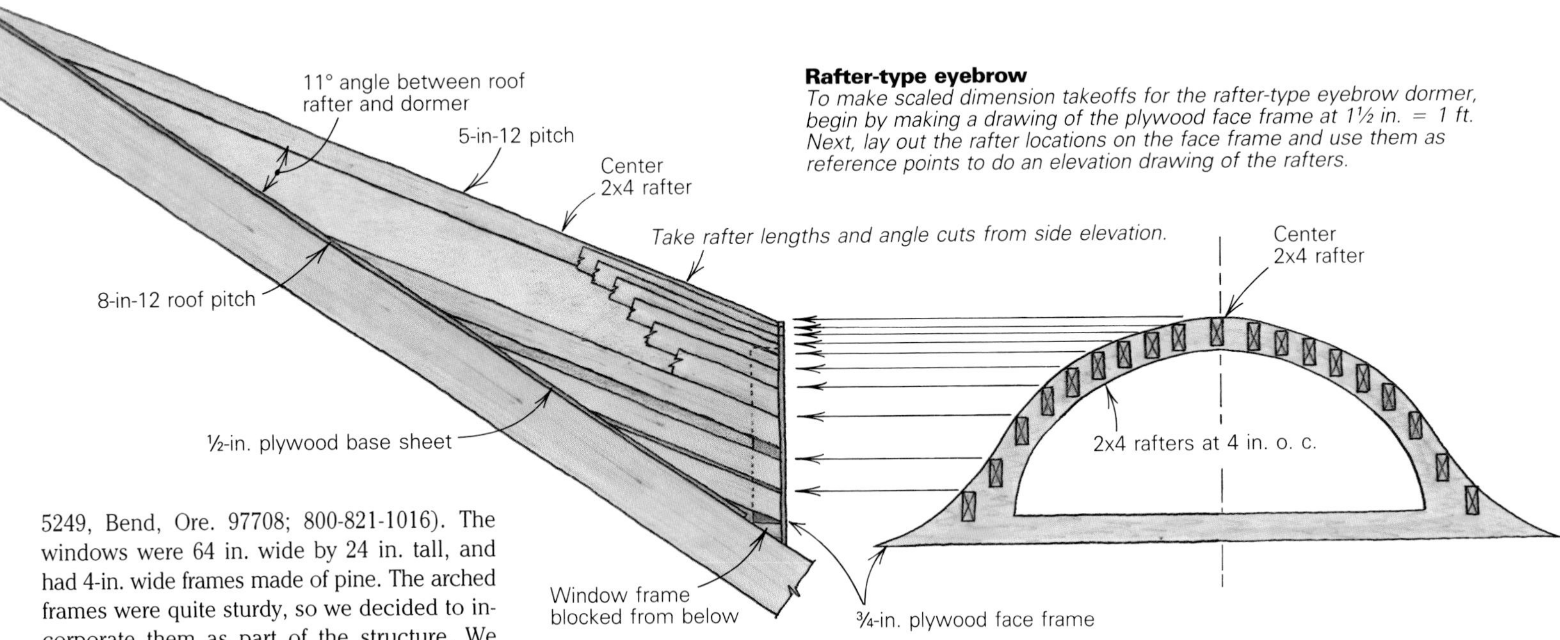

Rafter-type eyebrow

To make scaled dimension takeoffs for the rafter-type eyebrow dormer, begin by making a drawing of the plywood face frame at 1½ in. = 1 ft. Next, lay out the rafter locations on the face frame and use them as reference points to do an elevation drawing of the rafters.

5249, Bend, Ore. 97708; 800-821-1016). The windows were 64 in. wide by 24 in. tall, and had 4-in. wide frames made of pine. The arched frames were quite sturdy, so we decided to incorporate them as part of the structure. We faced them with ¾-in. ACX plywood, which would serve as a vertical surface for attaching the rafters, as well as backing for a stucco finish (top photo, right). At the top of the arch, this plywood face frame is 3¾ in. wider than the window frame. This dimension accommodates the 22½° plumb cut of a 2x4 rafter on a 5-in-12 slope—the pitch of our dormer. I screwed the face frame to the first window; then I braced it firmly on the roof, exactly on its layout between the roof trusses.

An arched window 2 ft. high with a base about 5 ft. wide makes a pretty tight curve for the dormer roof sheathing to follow. To make sure the curves stayed smooth and to ensure plenty of backing for the plywood, I decided to put my 2x4 rafters on 4-in. centers. I laid out their centerlines on the base of the window, and then used a level held plumb to transfer them to the arched portion of the face frame. Next I got out the string.

The luxury of full scale—Normally I make detailed drawings of unusual framing assemblies to familiarize myself with the geometry involved while still sitting on terra firma. This first dormer proved to be an exception to that rule, as I had the luxury of mocking it up on the garage roof. Still, clambering around on an 8-in-12 roof deck isn't everybody's idea of fun, so I would recommend doing a detailed drawing of the dormer's essential components, and then using it to scale the lengths and angles (more on this in a minute).

The opening in the roof made by an eyebrow dormer is bell-shaped in plan (bottom photo, right), with the bottom of the bell corresponding to the base of the window. Finding the shape of the bell became the next task.

First I tacked a couple of sheets of ½-in. ACX plywood to the roof deck (one over the other) so that their right edges were aligned with the centerline of the dormer. I specified plywood with an A side for all the plywood parts of the dormers (except for the sheathing) because it's much easier to draw accurate layout lines on a smooth, knot-free expanse of plywood than it is on a bumpy C or D side.

Then I cut a 2-ft. length of 2x4 with a 5-in-12

Plywood face frames screwed to arched window frames support the ends of the 2x4 rafters. The rafters are on 4-in. centers, and are parallel to one another. Note the uphill ends of the rafters. Their outside and inside corners touch the base and determine its shape.

Before the window and the rafters are installed, a bell-shaped base sheet is affixed to the roof deck. The next step will be to cut away the decking inside the base.

Drawings: Christopher Clapp

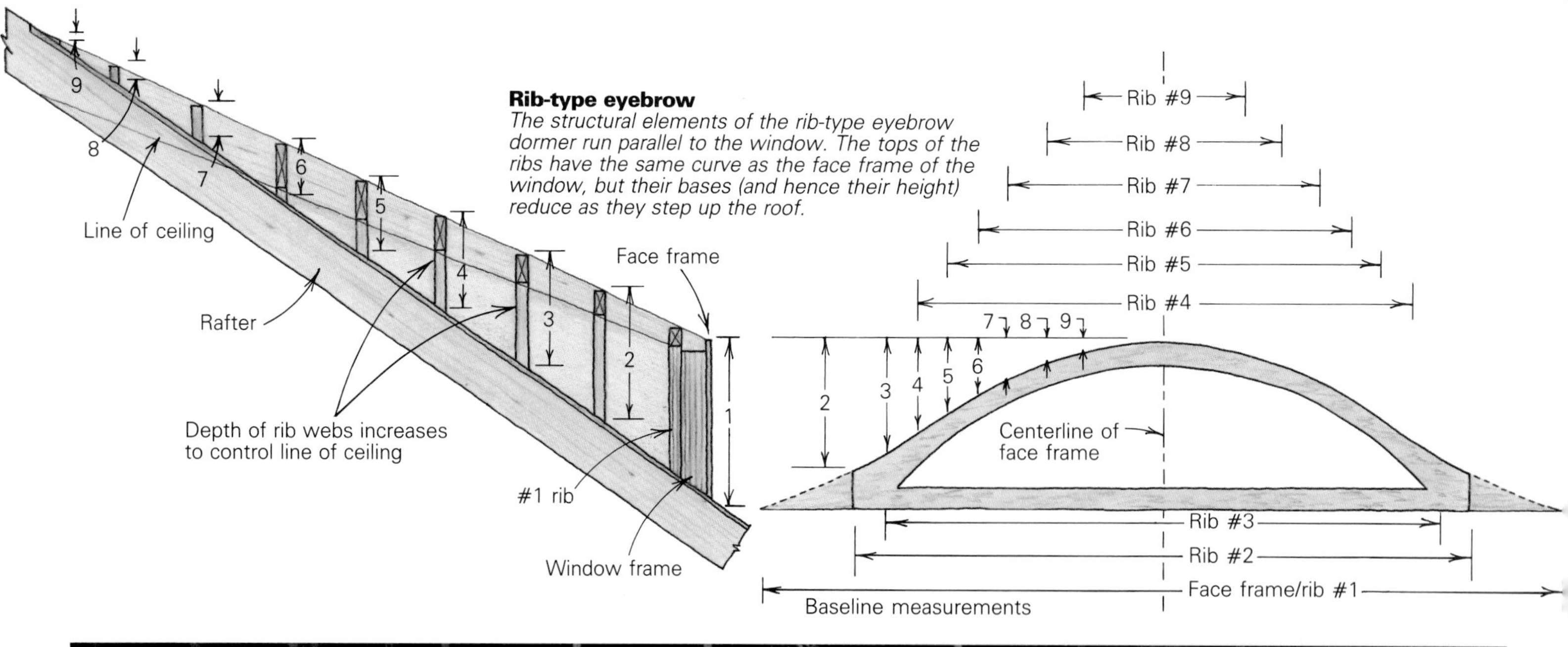

Rather than adjust the ribs on the roof, the author brought the roof to the workspace. Temporarily tacked on this 8-in-12 worktable, the corners of the rib bases are noted on the plywood to generate the bell-shaped base. The paper pattern will be used to cut out the plywood sheathing.

plumb cut on one end to act as a dummy rafter and rested the cut end on the top of the window frame. I held a string flush to the top edge of the dummy rafter, using a helper to hold it at the uphill end. This stringline represented the top of the center rafter, and its intersection at the roof was marked. Then the stringline-and-mark process was repeated for each rafter positioned to the left of center. By connecting the marks with a smooth curve, I had the outside line of the bell. To find the inside line of the bell, I measured the angle between the string and the roof deck. This angle (11°) represented the cut needed on the bottom of the dormer rafters where they intersect the garage roof. Because all the rafters are at the same pitch and parallel to one another, this cut is the same for all the rafters. By making a sample cut on a short length of 2x4, I simply placed the tapered end on each rafter layout line and marked the inside corners. Connecting the dots gave me the inside of the bell curve. Because the rafter layouts are symmetrical, the bell-shaped base for half of the dormer is the mirror image of the other half. Therefore, half of one base is all the layout template needed for one dormer. The same applies to the rafters. Once I had them measured and cut for half of the first dormer, I had templates for all the rest, allowing me to cut the parts quickly for all five dormers.

All this string-holding worked okay. But if I were to do this again, I'd do a drawing at a scale of 1½ in. to 1 ft. showing the elevation of the window from two vantage points (drawing previous page). At this scale, it would be easy to make dimension and angle takeoffs for the

From *Fine Homebuilding* magazine (February 1991) 65:80-84

lengths of the rafters and the angle at which they intersect the roof and the face frame. Then I'd use the rafters instead of the stringlines to figure out the bell shape of the plywood plate.

Assembly and sheathing—Once the plywood plates were cut out, we screwed them to the roof deck with galvanized drywall screws. Then we cut out the decking on the interior side of the base sheet, and screwed the eyebrow rafters to the plate and the face frame.

When one of the dormers had all its rafters, we draped 30-lb. felt over half of it and trimmed the felt along the valley formed by the intersection of the garage roof and the side of the dormer. This gave us the pattern we needed for marking cut lines on the ⅜-in. CDX plywood sheathing.

Bending a sheet of plywood over a radius this tight while holding it on the layout can be daunting—especially on a steep roof—so I wanted to prebend the largest pieces. Shallow kerfs on the underside of the plywood would have allowed it to bend more easily, but the structure of the dormers is visible from below so I wanted to avoid kerfs. Instead, I made a simple bending form out of a sheet of plywood with some 2x4 cleats nailed to the long edges. Then I stuffed several of ⅜-in. plywood sheets between the cleats, soaking each one liberally with the garden hose. Left in the form for a couple of days, the sheets took on a distinct curve, making them easier to bend over the rafters. Each dormer has one layer of ⅜-in. plywood affixed to the framing with 1-in. staples. We left the tops of the rafters unbeveled, but added beveled strips for better bearing where the sheets abut one another.

The garage roof had skip sheathing atop its decking to give the cedar shingles some breathing space. We carried the skip sheathing over each dormer by stapling a double layer of 6-in. wide redwood benderboard (5⁄16 in. thick) on top of the plywood. I had wondered what kind of valley flashing would be needed at the junction of dormer and roof. As it turned out, we didn't need any. The roofers wove layers of shingles together with very little exposure to form the valleys (bottom photo, p. 90).

After the rafters, the ribs—Unlike the multiple eyebrows on the garage roof, the single eyebrow atop the pool house had to have a finished ceiling underneath it. I decided that this extra wrinkle warranted another approach to the eyebrow's structure. Granted, you could hang blocking and furring strips from the bottom of a rafter-framed eyebrow to make a smooth transition from a flat ceiling plane to one with an arch, but why not make the bottom of the eyebrow structure conform as closely as possible to the shape of the arched portion of the ceiling? To that end, I worked up a full-scale drawing of the dormer (drawing facing page) on the recreation-room floor.

The window hadn't yet been ordered for this eyebrow, allowing me to design the arch from scratch. I made it long and low, taking the bulk of its face frame from a 10-ft. sheet of ¾-

The bell-shaped base of the rib dormer rests on rafters that will soon be headed off and trimmed back (top photo). In the bottom photo you can see how the two rafters at the top, now cantilevered over a new ridge beam, have been cut back at a taper to keep them out of the arched ceiling plane. To their right, an angled doubler picks up the weight of the ribs bearing on the base sheet. Benderboard strips backed by 2x blocking define the curve of the arched ceiling.

in. plywood. The short reverse-curve valley returns at each end were made of scabbed-on pieces of plywood (photo, p. 92).

The ribs are on 16-in. centers, and their bottoms are cut at an 8-in-12 pitch to match the roof slope (drawing, p. 92). As the individual ribs step up the roof, their overall depth decreases along with their width. Meanwhile, their arc at the top remains the same as that of the face frame. By taking direct measurements off the full-scale drawing, I got the overall depth of each rib. Then I measured down on the centerline of the window face frame to find the perpendicular baseline to read the width of each rib. To add a little extra complexity to the project, I had to increase the depth of the web of each rib in a sequential manner. This allowed the arched portion of the ceiling to make its transition into the cathedral ceiling without crowding the ridge (photo below).

I made all the parts for the rib-type dormer out of ¾-in. plywood. The face frame is a single layer, the built-up window frame has 5 layers and each rib has 3 layers. Rib number 1 is screwed to the back of the window frame.

Working with a full-scale drawing made for accurate and speedy work. But the pieces were large and cumbersome, and temporarily tacking them to the roof to figure out the shape of the base didn't sound like any fun at all. I probably could have used the full-scale elevation drawing to extrapolate its shape, but whenever I have to deal with unusual concepts, like sections of cones on inclined planes, I take comfort in three-dimensional models.

While regarding the cavernous interior of the recreation room, a solution occurred to me. Why not build a mockup of the roof, with one end of the rafters firmly planted on the recreation-room floor? Within an hour, I had a fake roof in place. I used the plywood that would eventually become the base sheet for its sheathing.

As each rib was cut out according to direct measurement takeoffs, I tacked its base to the fake roof and braced it plumb with a temporary alignment spine (photo, p. 92). Once I had all the ribs tacked to the mockup, I marked the inside and outside points where their bases engaged the plywood. These points gave me the reference marks I needed to make the bell-shaped base for the ribs. After taking the ribs down, I drove 8d nails at each mark, leaving enough of the nails exposed to act as stops. Then I used a ¼-in. by 1½-in. strip of straight-grained redwood benderboard held against the protruding nails to generate the curve for the base sheet.

The base sheet tucks into a bay between a pair of new timber-framed trusses (top photo, previous page). Unlike the installation of the rafter-type dormers, this one went into a roof that hadn't been planned with an eyebrow dormer in mind. This meant that some rafters had to be removed, and their loads picked up and transferred to new structural members.

A plaster finish on the arched ceiling flows into the drywall covering the plane of the rafters.

Blocking and benderboard—Before I took apart the mocked-up structure of the dormer, I made a rosin-paper pattern to guide the cutting of the plywood sheathing. Like the pattern for the dormers on the garage, this one could be flopped to be the pattern for the other side of the dormer.

Assembling the ribs began from the bottom up. With the face frame and its accompanying arches firmly attached to the base sheet and diagonally braced plumb, all the succeeding ribs were quickly placed on their layout marks. They were then decked with a single layer of ½-in. plywood. As before, I built up skip sheathing over the top of the eyebrow with two layers of 5⁄16-in. by 6-in. benderboard.

Picking up the loads of the removed rafters and carrying the curves of the ribs into the plane of the ceiling was the next task. Our engineer recommended a couple of doubled 2x8s as support for the legs of the base sheet. These doublers run diagonally from the ridge beam to the top chords of the new timber-framed trusses (bottom photo, previous page). This photo also shows the finicky blocking that it took to pick up the unsupported edges of plywood sheathing and to carry the arc of the eyebrow into plane with the rafters. Benderboard was also useful for this task. In places, I was able to extend the curve from a rib to the rafters with a strip of benderboard and then fill in the remaining gaps with solid pieces of blocking shaped to fit.

The ends of the benderboard abut the edges of the drywall that cover the flat parts of the ceiling. At the transition to the curve, expanded metal lath was stapled over the benderboard, and the junction between the flat ceiling and the eyebrow's arch feathered with plaster to make an invisible seam. □

James Docker is a building designer and general contractor living in San Carlos, California. Photos by author except where noted.

Framing for Garage Doors

Think about the door before you pour

By Steve Riley

The low hum and clickety-clack rhythm of an automatic garage-door opener is a sound of our times. And yet the ingenious hardware and signal receivers that operate the doors are affixed to the most basic form in building: a beam supported by posts or, as we call them today, a header on trimmers and king studs. Making sure the framing around the door is assembled accurately and with structural integrity is the topic of this article. Building a garage and getting it ready for a door typically involves several different trades, and it takes some planning to make sure the guy who finally installs the door can do a clean job with no complaints. After all, nobody likes to chip away concrete to make room for the door tracks, and nobody wants to pay for a custom garage door because the stock one didn't quite fit in the opening.

Door types and sizes—Stock residential garage doors generally range in size from 8 ft. wide and 6 ft. 6 in. tall to 18 ft. by 8 ft. A good set of construction drawings will call out the garage-door details on the framing plans and on the elevation sheet. If the drawings lack this information, I check with the clients to find out what kind of vehicles they plan to park in their garage. For a midsized car, the typical door will have to be 8 ft. wide and 7 ft. tall. But most people want the flexibility of pulling a Suburban with a luggage rack into their garage, which requires a 9-ft. wide door that is 8 ft. tall. The extra width also reduces the chance of a driver ripping off the passenger-side mirror.

Here in the Wood River Valley in Idaho where I build houses, clients prefer sectional garage doors. They are made of four or more horizontal panels linked by hinges, with rollers that ride in tracks mounted to 2x6 side casings (drawing below right). The hardware that mounts over the door is attached to a 2x12 head casing.

Sectional doors don't mount between the jambs. Instead, they mount just inside the garage and are sealed against the weather by stops nailed to the jambs. The weather seal is one reason why folks around here prefer sectional doors. For an even better seal, the stops can include weatherstripping.

In some parts of the country, people still use one-piece, or tilt-up, garage doors (drawing middle right). These are less expensive than sectional doors and easier for builders, or homeowners, to install. One-piece doors fit between 2x side jambs that extend into the garage 2 in. beyond

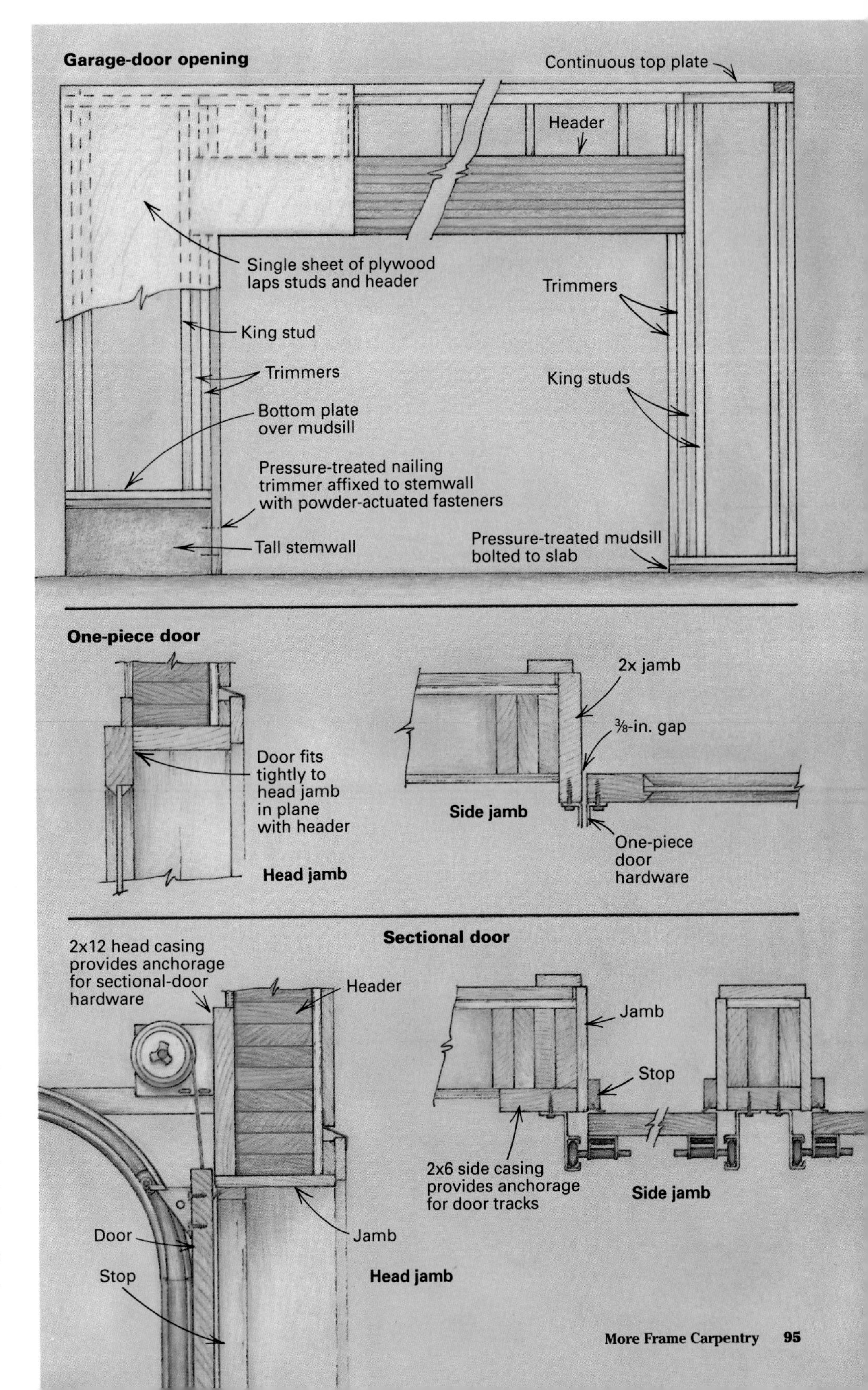

Drawings: Christopher Clapp

Framing with steel

By Tom Conerly

Building designers in Santa Cruz, California, have their share of specialized design criteria. And sometimes the criteria can work to cross purposes. Take, for example, two of the conditions I had to reconcile in the design of a duplex for one of our historic residential districts.

The original duplex was destroyed in the 1989 earthquake, along with many of the neighboring homes and commercial buildings.

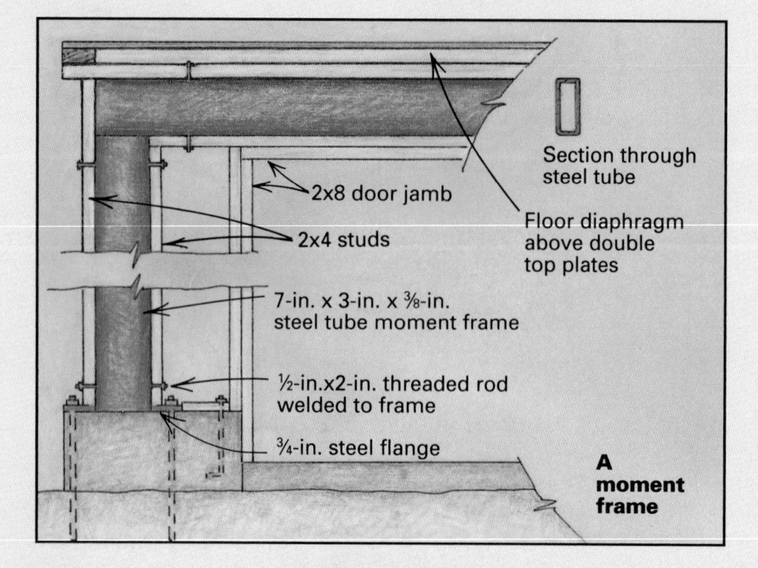

A moment frame

Fortunately, enough of the original structures remained undamaged to preserve the flavor of the old part of town, and to make sure it stays that way, any new construction on this street has to be compatible with the neighborhood. So among other guidelines, the two single-car garages required for the new duplex had to be kept separate to reduce their bulk. I put the living spaces over the garages, which fits with the row-house look along this street. But stacking a floor over a narrow garage that's mostly doorway makes for some potentially heavy loading on the garage-door framing.

As you can imagine, we have to adhere to some of the toughest seismic design standards anywhere in the world. Ordinarily, I would reinforce the corners of the garage-door wall with plywood sidewalls. But our tall, narrow building precluded that option because the garage sidewalls were too narrow to contribute much stiffness. Fortunately, our project engineer, Michael Martin, had a budget-conscious alternative for us: the moment frame (photos below)

Martin learned about steel-moment frames designing fast-food restaurants, a job that required him to reinforce openings for large windows in wood-frame buildings. A moment frame relies on a stiff connection between framing members to resist the *moment,* or forces that cause bending, in a structural member. It's difficult to make a moment connection with wood because the fasteners tend to act as pins, thereby allowing some flex. No flex is allowed in a moment connection.

As shown in the drawing at left, our moment frames are made of 3-in. by 7-in. by ⅜-in. thick steel tubing. They are bolted to the foundation with ¾-in. thick steel flanges over 1-in. dia. threaded steel rods (photo below). The rods are 27 in. long, and they have nuts and washers on their lower ends to anchor them in the concrete. The 2x wood framing members that sandwich the steel are secured by 2-in. long pieces of ½-in. threaded rod welded to the sides of the steel tubes.

We had the two frames fabricated by a local welding shop for $650 apiece, plus another $100 to transport them to the site. To ensure accuracy, our contractor, Rob Moeller, provided the shop with flange templates that gave the exact positions of the foundation bolts.

We scheduled the arrival of the frames to coincide with a crane that had to be there anyway to lift other materials onto the roof. It took an hour's worth of crane time to place the two frames and another couple of hours of carpentry to bolt the 2x stock to the steel.

I put narrow gable roofs on top of each garage to break up the façades of the duplex and to keep them from looking top-heavy. As a consequence, the upstairs walls are about 2 ft. back from the plane of the moment frame. Shear loads are transferred from the walls to the frame by way of a horizontal plywood diaphragm (for more on how wood-frame building react in earthquakes, see the article on pp. 58-63).

— Tom Conerly is a building designer based in Santa Cruz, Calif. Photos by the author.

From *Fine Homebuilding* magazine (April 1992) 74:57-59

the plane of the header. The hardware is bolted to the edge of the jamb. Many builders make the mistake of using redwood jambs for this detail. Garage-door installers prefer Douglas-fir jambs because lag bolts hold better in them. Unlike sectional garage doors, one-piece doors require a ⅜-in. gap at each side to keep them from scraping the jambs during operation.

The type of door and the hardware it uses influence the ceiling height of the garage. The minimum overhead clearance for standard sectional-door hardware is 12 in. above the bottom of the head jamb, plus another 2 in. if the installation includes an automatic door opener. Reduced-clearance tracks that fit in a 6-in. deep space can be specially ordered from most manufacturers for low-headroom situations.

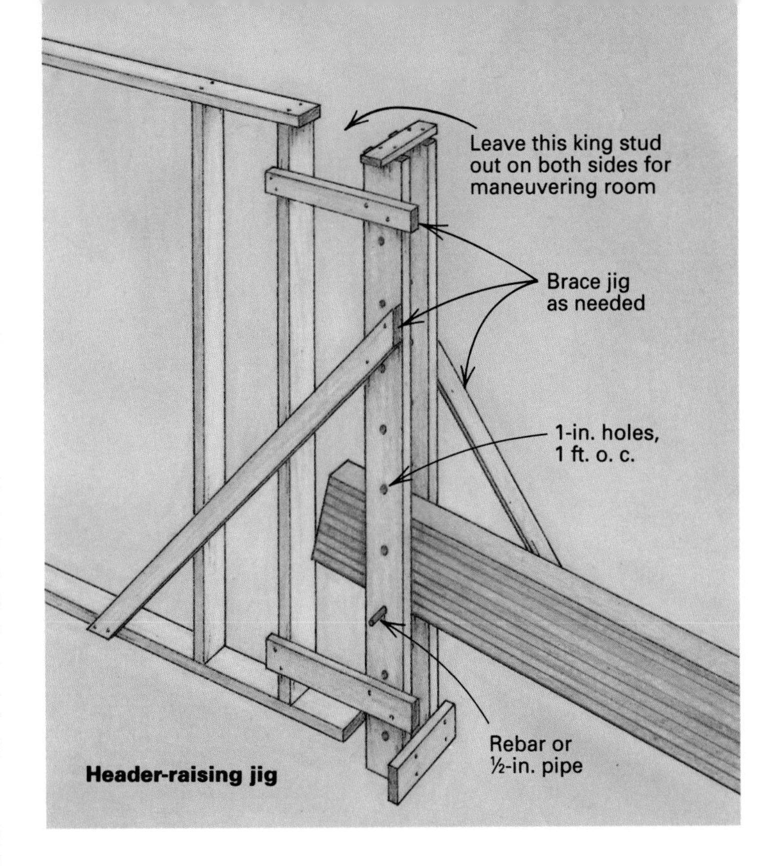

Header-raising jig

Stemwall or slab—Garage walls bear on either a monolithic slab that includes a footing or on stemwalls that border a slab. In either case, make sure anchor bolts and hold-downs are accurately placed—especially in the short walls that flank the garage opening. These walls are typically 2 ft. long and require a pair of anchor bolts, each within 1 ft. of the ends. Nowadays, building codes often require metal hold-downs as well. To fit all the hardware, I make these short walls as long as possible by running them past any adjoining walls instead of abutting them.

Of the two foundations, I prefer the stemwall because it's easier to position the anchor bolts on the forms prior to the pour. If hold-downs are called for, consider the strap-type, such as Simpson's HPAHD22 (Simpson Strong-Tie Co., Inc., 1450 Doolittle Dr., San Leandro, Calif. 94557; 510-562-7775) if it satisfies the engineer. This hold-down has a hook that angles into the stemwall while the strap extends up and is anchored to the side of the studs. It's easier to deal with than hold-downs that require foundation bolts.

For a sectional door, I want an opening between stemwalls that is equal to the width of the door plus the thickness of two nailing trimmers and two jambs. For a one-piece door, the opening between the stemwalls is equal to the width of the door plus two jambs and a ¾-in. gap. I don't leave a shim space because I make sure my trimmers are plumb during framing.

King studs, trimmers and headers—Whether on a slab or on a stemwall, I prefer that the ends of my garage headers rest on two trimmers (top drawing, p. 95). With a tall stemwall (8 in. or more), I add a third pressure-treated trimmer that extends all the way to the garage floor to provide a nailing base for the bottom of the jamb and casing. I secure this trimmer to the stemwall with a couple of powder-actuated fasteners.

For a sectional door, I want at least four framing members on each side of the door opening. That way I've got backing for the drywall and for the 2x6 casings that carry the tracks for the doors. The framing members can be two trimmers and two king studs or three trimmers and one king stud. When framing two, one-car sectional doors next to one another, you need a column at least 6½ in. wide between them for the door tracks (bottom drawing, p. 95).

Although the front of a garage is a continuous wall, the large opening makes it tough to erect this way. So I frame the short sections that flank the door as separate walls and connect them later with a continuous top plate. I frame these short walls with bottom plates drilled to fit over the anchor bolts holding down the mudsills. After lifting a wall into place, I make sure the outside edge of the bottom plate is flush with the face of the stemwall. This step ensures that the sheathing and the siding will be outside the plane of the concrete. At this stage, stretch a stringline from corner to corner to make sure the walls are in the same plane—stemwalls aren't always exactly where they're supposed to be.

Bottom plates aligned, I nail them to the mudsills. Then I nail the outside corners of the adjoining walls and brace the narrow walls so that they stay put as we place the header.

***Custom veneer.* A sectional garage door can blend with the house when compatible siding is affixed to it. Horizontal finishes (lap siding or shingles) can make the lines between panels virtually disappear. Photo by Charles Miller.**

We can muscle a single-door header into place by hand (for example, a 9-ft., 6x14 weighs about 175 lb.). But it's almost essential to have a crane or a forklift on site to place a header large enough to span a double-wide door. Lacking a crane or a lift, we resort to a ladder-like header-raising jig that allows us to lift one end of the header at a time (drawing left). Using this jig requires the cripples and the two king studs to be left out of the walls for maneuvering room until the header is in place. I stick with either glulams or laminated veneer beams for header stock because large, sawn timbers have a tendency to twist and check. Also, I don't use built-up headers made of 2x stock and plywood because the labor required to fabricate them costs more than the extra expense for solid stock, and built-up headers are rarely as accurate as glulams or veneered beams.

Once the header is positioned, I tie it into the wall by nailing through the king stud into its ends—a typical nailing schedule is pairs of nails on 2-in. centers. Then I add my cripples and double plates to tie it all together. After plumbing and stringlining, the wall is ready for plywood. It's important to lap full sheets over the intersections of the header, the trimmers and the king studs to reinforce these areas (top drawing, p. 95). Now I can put up my 2x casings for the door hardware and call in the door man.

Because carpenters do not install garage doors regularly, it's best to have an expert install the door. Otherwise, count on watching two or three carpenters spend several hours hunched over a set of instructions while they figure them out.

A custom touch—The most economical garage doors have hollow-core panels skinned with metal or hardboard. The next step up is the same doors, but insulated. High-end stock doors are the raised-panel variety made of solid wood.

An elegant way to make a garage door match its surroundings is to skin it with the same siding that covers the house (photo below). This step adds some weight and, in my experience, works best on doors that are 9 ft. wide or less. Consult with your door installer to find out how much weight the hardware can bear.

For a substrate, I specify insulated hardboard doors with a smooth face. Then I use galvanized staples and construction adhesive to apply the veneer to the door sections. I've put siding on doors before installation, and I've sided them after installation. The second method is better because the siding adds weight to the door, making it more difficult to install. Also, the alignment of the siding stays true when affixed to a door that's in place. The installer will come back to adjust the spring tension to offset the extra weight. □

Steve Riley is a general contractor based in Ketchum, Ida.

Composition Panels

OSB, particleboard, MDF and hardboard: what they are and what they can do

by Charles Wardell

Quality wood is becoming hard to get. Around the world, forests are disappearing; in the U. S., many once-vast stands of old-growth timber have been replaced by second- or third-growth trees of smaller size and lower quality. The environmental consequences of this dwindling supply is the topic of much heated debate, but the effect on the builder is certain: higher prices and lower quality. Just ask anyone who's tried to find a reasonably priced supply of straight 2x4s lately. To combat this problem and to help keep itself afloat, the forest-products industry has been scrambling to develop products that use wood more efficiently. The result has been an expanding array of wood-based composites that are manufactured by taking small trees apart and putting them back together in more efficient forms. A full catalog of these would fill a small volume. Most, however, are variations on the four major composition panels (photo right): oriented strand board (OSB), particleboard, medium density fiberboard (MDF) and hardboard. OSB is considered a direct replacement for structural plywood, while the other three are meant to be used in less demanding applications. In this article I'll give an overview of these four products, along with some tips on the right ways to use them. If you're familiar with these, you'll have a good basis for evaluating any other composites that come down the line.

The major composition panels. **Left to right: oriented strand board, particleboard, medium density fiberboard and hardboard. Far right: OSB is available square-edged or T&G. Most OSB is treated with an edge seal (red here, but color varies by manufacturer).** ***Photo by Robert Marsala.***

Particles and adhesives—Composition panels are made by mixing wood particles or fibers (photo next page) with a synthetic adhesive and then subjecting them to heat and pressure (see sidebar, p. 100). Unlike plywood, which has been around for centuries (forms of it have been found in Egyptian tombs, for instance), composition panels are creatures of this century, developed in response to specific market forces. The story of these products is largely the story of adhesives. Advances in adhesive technology over the past 60 years have made possible a wide array of composites. Most adhesives are 50% to 100% polymer. A polymer is a large, complex molecule made up of repeating chains of smaller molecules. In an adhesive it's the stuff that forms the actual bond. The rest of the adhesive usually consists of a carrying agent. Phenolic-formaldehyde and urea-formaldehyde resins are the most common adhesives. Both use formaldehyde as a cross-linking agent; the formaldehyde molecules act like the couplings on a freight train, with individual phenol or urea molecules representing the box cars. Phenolic resins are waterproof, but more expensive than urea-formaldehyde. The phenolics are the resin of choice for most structural-use panels. Urea-formaldehyde is classified as water-resistant, not waterproof. Urea-bonded panels are meant for less demanding applications, like cabinet carcases and furniture.

The other half of the equation is the particle. Varying the amount and type of particles and adhesives lets the manufacturer engineer in any one of a number of properties. Like any composite, a composition panel can be made stronger, more consistent or more predictable than a comparable piece of solid wood. Its exact properties will vary with its intended end use. The small particles used in interior panels yield a smooth surface that's ideal as a substrate for paints or laminates. But a structural panel made from large flakes, while not as smooth or uniform, will be stronger and more resistant to swelling.

Oriented strand board—OSB has been around since the early 1980s. A lot of people confuse it with waferboard. From a structural perspective, there's not a big difference between the two products, so what I say about OSB will apply to waferboard also. Both products are designed for structural use and are performance rated by the American Plywood Association (APA), the main trade association for the structural-panel industry. As their

From *Fine Homebuilding* magazine (April 1991) 47:77-81

names imply, waferboard is made with randomly aligned wafers, OSB with directionally oriented strands. The result is that OSB, like plywood, is strongest in the long dimension, while waferboard is equally strong in all directions. One of the great advantages of OSB is that it uses less material than waferboard: for a given span rating, a sheet of waferboard will be thicker than its OSB counterpart. Because of this, OSB has replaced waferboard for most structural applications.

OSB is used mainly as roof and wall sheathing, as subflooring and as a combination subfloor/underlayment (APA-rated Sturd-I-Floor). Other OSB products include siding and wood I-beam webs. OSB isn't a product suited for long-term exposure to the elements, however; it's rated Exposure 1 (for normal construction delays). Because of this, OSB siding is made with a weather-resistant, phenolic resin-impregnated overlay that can be coated with any exterior paint or opaque stain. Like plywood, OSB is available in a Structural 1 grade with 10% more shear strength than ordinary sheathing. This product is used mostly in engineered applications, where the structure may encounter high wind loads or earthquake conditions.

There are a couple of things you *can't* do to OSB yet. These include preservative treating and fire-retardant treating. And though some companies make ¼-in. OSB underlayment, the APA hasn't developed a standard for it (though they're working on it).

The material can be worked with normal tools, though carbide blades are recommended for prolonged cutting. OSB can be glued with an elastomeric construction adhesive, or with any glue that's meant to be used with wood. Nails can be driven as close as ¼ in. from panel edges without breaking out.

Subflooring and Sturd-I-Floor should be installed over a dry, ventilated space and allowed to acclimate before installation. When using square-edge panels without an underlayment, you'll need to block around the edges. No blocking is needed for T&G Sturd-I-Floor. Although Sturd-I-Floor is made so that finish flooring can be laid directly over it, there's one notable exception. Most vinyl-flooring manufacturers recommend using an additional panel underlayment beneath vinyl flooring. Use adhesives and underlayment recommended by the flooring manufacturer. Some people believe the edge seal used on OSB underlayment will bleed through vinyl floors, especially thin vinyl. It's a controversial issue, but a number of claims have been filed against manufacturers, and at least one has stopped marketing OSB underlayment. Finally, OSB shouldn't be used as a substrate for ceramic tile; when installing tile over an OSB subfloor, lay down a plywood underlayment first.

Proper spacing is crucial with OSB. The APA recommends that panels be gapped ⅛ in. at the ends and ¼ in. along the edges. To accommodate this, manufacturers may undercut their panels up to ⅛ in., although ¹⁄₁₆ in. is more common. A 10d box nail may be used to gauge the spacing. Panel edge clips are also available at most suppliers. In addition to acting as spacers, these clips help stiffen unsupported panel edges. They're used mostly with roof sheathing, where the structure must be able to support heavy snow loads, as well as the weight of workers. John Rose, an engineer at the APA, told me that roof or wall sheathing should not be glued to framing. If a glued panel gets wet during construction, it can buckle. Subflooring and Sturd-I-Floor are thicker and are installed on shorter spans. Because of this there's less tendency for them to buckle, and they can be safely glued to the floor joists.

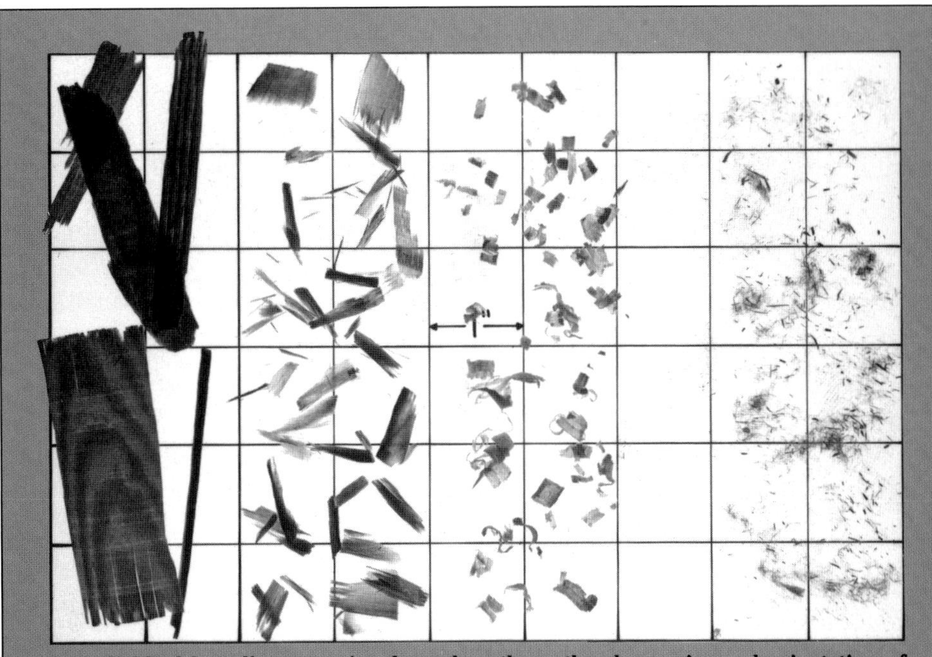

An individual board's properties depend partly on the shape, size and orientation of the individual particles or fibers. From left to right: wafers, ring-cut flakes, planer shavings and pressure-refined fibers. Each grid square equals 1 sq. in. *Photo courtesy of Washington State University.*

OSB and moisture—One of the biggest complaints builders make about OSB is its susceptibility to thickness swelling. Although plywood will swell, too, OSB is made with a lower moisture content (generally 2% to 5% as opposed to around 6% for plywood), and so must take on more moisture to reach equilibrium. This means that OSB should be handled a little differently than plywood. For example most manufacturers treat their panels with an edge seal that will prevent moisture from wicking in. But if a sheet is dropped on edge or hit with a hammer, the seal can be broken. The panel will then be more likely to soak up moisture and to swell along the edges.

The product should be covered up as soon as possible after it arrives on the job site. It's also a good idea to cut the steel banding on panel bundles right after delivery to prevent edge damage. Bundles should be stacked level on 4x4s or doubled 2x4s and covered with a plastic tarp. The tarp's cover should be pulled away from sides and edges of the bundle to provide ventilation and to prevent moisture and mold formation.

One builder in Tennessee told me that because he needs to cover OSB so quickly, he won't use it as subflooring. He lays the subfloor before beginning the wall framing—often months before the house is dried in—so there's a great chance of moisture damage.

Another builder I talked to takes a different approach. Bill Eich builds in Spirit Lake, Iowa. He installs OSB subflooring or Sturd-I-Floor, then immediately applies a water repellent. Even so, there's often swelling along the edges of the OSB by the time the roof is on. This is easily cured by hitting the panel edges with a floor or belt sander. Eich also thinks that it's best to let OSB roof sheathing get a wetting before it's covered. The material is bound to pick up moisture, either from construction delays, from the framing, or from moisture in the air. He figures that letting it do so before the roof is on decreases the chance that any thickness swelling along the edges will be telegraphed to the shingles. I ran Eich's method past an APA engineer, who agreed that it's sound—as long as the sheathing isn't left exposed so long that the edges begin to swell.

Tests conducted at the APA and at the Forest Products Laboratory in Madison, Wisconsin, have rated OSB's nail-holding ability as good as, or better than, an equivalent thickness of plywood. But not everyone agrees with these results. Don Bollinger, a hardwood-flooring contractor from Seattle, Washington, refuses to guarantee his floors when they're installed over OSB. In fact, complaints from floor installers have led the National Oak Flooring Manufacturers Association (NOFMA, P. O. Box 3009, Memphis, Tenn. 38173-0009; 901-526-5016) to recommend that oak flooring not be installed over OSB subfloors. A representative of that association told me that although new OSB seemed to hold nails as well as plywood, there were some questions about its performance after exposure to water (as it would be during a construction delay) or high humidity. To help answer these questions, NOFMA has agreed to sponsor tests at Virginia Tech comparing the nail-holding abilities of OSB, plywood and board subfloors under various moisture conditions. A lot of builders, including Bollinger, believe that you shouldn't use OSB anyplace where it could get soaked from a leak in the plumbing system. Bollinger would include all kitchens and bathrooms in this category. "You have to look at the long-term likelihood of moisture trauma" he says, "such as a plumbing burst when the owners are away. With OSB, one trauma often means that you have to rip up the whole floor."

Particleboard—Particleboard is a urea-bonded panel that's made for interior applications. It was first developed in the 1930s as a way to use planer shavings, sawdust and other waste materials. Most American particleboard is used for underlayment, cabinet carcases and furniture. The American National Standard for particleboard (ANSI 208.1) includes a phenol-bonded structural grade, but 98% of what's manufactured in this country is of the nonstructural, urea-bonded variety. Particleboard comes in ten grades and three density ranges (high, medium and low). The particleboard you see at the lumberyard is usually medium density. In general, denser panels are heavier but stronger. They also have smoother, tighter edges than low-density panels. Two special grades are produced for manufactured home decking and for underlayment.

Particleboard isn't known for its nail-holding ability. Ring-shank nails can be used to fasten underlayment, but for most nonglued applications, screws are the fasteners of choice. Almost any open-thread screw is acceptable, but for better holding power use particleboard screws. These resemble drywall screws and can be driven with a screw gun. The difference is that particleboard screws have higher profile threads with a deeper bite, as well as wider, stronger shanks for driving into dense material. Because particleboard's surface is so dense, screw holes must be countersunk before the screws will seat flush with the surface. McFeely Hardwoods (712 12th St., P. O. Box 3, Lynchburg, Va. 24505; 800-443-7937) sells flathead steel particleboard screws with "nibs," or small cutting edges, formed into the underside of the screw head (photo next page). The nibs let the screw cut its own countersink. Particleboard is less dense at the center, so fasteners driven into the face hold better than those driven into the edges. When screwing into an edge, drill a pilot hole. This will help the screw hold better and will prevent deformation of the material.

Check your local building code before installing particleboard underlayment. As a general rule, however, the subfloor should be at least 18 in. above any crawl-space floors. Dirt crawl spaces should be covered with a vapor barrier having a maximum rating of 1 perm.

Reinventing the tree

To learn more about what composition panels are all about, I spent a day last August at Georgia-Pacific's Woodland, Maine, OSB plant. The Woodland facility is a study in contrasts. It's one of the oldest plants in continuous operation in the country. The cavernous steel building's hot, humid air is thick with dust and with the deafening sounds of machinery. At the same time, it's largely an automated operation. Every step in the complex process of turning trees into 4x8 sheets of sheathing and flooring is monitored and controlled by three central computers.

At one end of the plant, a continuous stream of trucks arrives carrying aspen logs with an average diameter of 6 in. to 8 in. (compare this to a plywood mill that can peel veneer down to only about 6 in.). Aspen is the wood of choice for OSB because it's easy to chip into large flakes. And though large flakes are harder to process than small ones, they make a stronger, more moisture-resistant board. Plant manager Ralph Hicks told me that Woodland needs about three-quarters of a cord of these logs to produce 1,000 sq. ft. of ⅞-in. OSB.

A crane lifts the logs onto a conveyer, which transports them to the slasher where they're cut into 8-ft. lengths. Next, they're run through a debarker, dumped into hot ponds and left to soak. Soaking makes the logs easier to flake, and the soaking time varies with the time of year; logs must soak longer in winter, when they need to thaw out as well as to soften. After soaking, the logs are recut into 33-in. lengths and fed to a machine called a flaker.

The flaker spits out flakes that measure from 0.025 to 0.030 in. thick and anywhere from toothpick size to 3½ in. wide. The mixture is then dried, and the wafers below ¼ in. wide are filtered out (the plant burns them as fuel). Board strength is partly a function of flake size. The larger the flakes, the stronger the board, so removing the smaller ones makes sense.

The flakes are then blended with a powdered phenolic resin. Some companies use an isocyanate adhesive to bind their OSB panels, but according to Hicks, isocyanates present hazards to plant workers. Because of this, Georgia-Pacific, like much of the industry, uses a phenolic resin to bind its OSB.

Once coated with adhesive, the flakes are sent to board formers where they're mechanically oriented into three layers, each perpendicular to the next. The face layers are oriented parallel with the 8-ft. dimension, the core with the 4-ft. dimension. The flakes exit the board formers as a continuous mat about 3-in. thick (photo below). The mat is cut into 8-ft. by 16-ft. sections and transferred to stainless-steel plates called screen cauls, named for their screenlike surface. The cauls are then loaded onto the press.

The press is to an OSB plant what an anvil is to a blacksmith—it's the tool that defines the business. Here, thousands of discrete wood flakes are re-formed into a product that's in some ways stronger than the trees they were cut from. Under 450° of dry heat, the mat is compressed to its finish thickness and the powdered resin liquefied. Inside each mat, the now-fluid resin flows into the wafers' microcavities, where it bonds with the individual wood molecules. The resin on the face of each board undergoes what Hicks refers to as "plasticization" from the heat of the press, leaving a thin surface film that will act as a moisture barrier. At the same time, the screen caul is impressed onto one face of the board, providing a slip-resistant surface for workers who install the product as roof sheathing. After the press, the boards are trimmed into 4x8 sheets and their edges coated with a sealer. They're then stacked, banded and loaded onto boxcars for shipment. —*C. W.*

Flakes exit the board formers as a thick, loose mat. They're then cut into sections and fed to the press. *Photo courtesy of American Plywood Association.*

Board subfloors should have at least a 1-in. nominal thickness and an 8-in. nominal width; panel subfloors should be at least ½ in thick, 15/32 for glue-nailed floors. Areas over furnaces and hot areas should be insulated and ventilated to prevent localized drying and shrinking. The best way to install particleboard underlayment is to glue-nail it. The National Particleboard Association recommends ordinary carpenter's glue applied with a paint roller. In this system, staples as well as nails may be used to fasten the floor. Particleboard has also been used with some success as a finish flooring material (see *FHB* #39, p. 43). It's cheaper than most other choices, dense enough to take traffic, and when finished, it's rich texture and color resemble cork.

Because it's a urea-bonded product that's sensitive to moisture, particleboard should only be installed after all concrete, plaster and lumber have dried to their approximate long-term levels. Store particleboard panels flat on bunks or skids in a dry place; it's best to have them delivered just before they're needed. From a materials standpoint, the best time to install underlayment is after completing the other interior finish work and just before installing the floor covering. The manufacturers recommend that particleboard underlayment not be used over concrete or below grade, in very wet areas such as bathrooms, or as a substrate for ceramic tile.

Similarly, particleboard underlayment or stair treads used near entries or in areas that are subject to wet traffic will need to be sealed with a moisture-resistant coating or covered with a waterproof surface like vinyl. Be aware, however, that some floor installers have expressed reservations about particleboard under thin resilient flooring because of telegraphing of seams—check with the manufacturer to see if particleboard is recommended as an underlayment for their products.

The biggest public concern about particleboard is offgassing of formaldehyde vapor. Manufacturers have reduced the amount of offgassing significantly over the last decade, but chemical sensitivity varies with each individual. To minimize offgassing, all exposed surfaces should be sealed; a clear polyurethane works well for this (for more on offgassing, see the sidebar on the facing page).

When urea-formaldehyde bonded particleboard is installed in a hot and damp environment, however (such as an habitually hot and steamy bathroom), it may continue to emit large amounts of formaldehyde vapor. Such an environment can cause hydrolysis (water-induced breakdown) of the cured urea-formaldehyde resin, resulting in release of formaldehyde vapors and loss of board strength. The process can apparently go on indefinitely if humidity and temperature remain at high levels. Use phenol-bonded exterior panels in these environments, instead.

Medium density fiberboard—MDF, like particleboard, is a urea-bonded panel designed for interior use. But while particleboard is made with discrete wood *particles* of various sizes, MDF manufacturers take the process a step further, breaking the wood down to its individual *fibers*. The uniformity of the fibers, along with the physical and chemical changes they undergo during the manufacturing process, guarantee a homogeneous panel with an extremely smooth surface.

The extra processing makes MDF more expensive than particleboard, but it also makes it easier to finish. It's main uses are for cabinetry, relieved door fronts, moldings and furniture. Expect to see more of it on the job site, however. MDF is especially useful where paintability is important. Its smoothness makes it an excellent substrate for thin laminates, and its uniformity allows sharp, clean edge-machining with minimal treatment before finishing. When high-quality carbide or diamond tools are used, MDF is easy to rout or saw, and there's no fuzzing or chip-out.

David Frane, a foreman for Thoughtforms Corp., a Boston-area construction company, says that his crews use a lot of MDF. One recent job called for a continuous shelf with an exposed edge several feet above floor height in a circular room. The shelf was to be painted, and using MDF eliminated the need for edge-banding. Frane also uses the product to make raised panels. It's stable, its edges can take any profile, and it finishes well. Because it comes in 4x8 sheets, large panels can usually be made in one piece. When gluing is necessary, Frane can use just about any adhesive made for wood.

The biggest complaint about MDF that I heard from builders concerned the dust. Because the material is made of fibers rather than larger particles, the dust is very fine. "It

	OSB	Particleboard	MDF	Hardboard
General uses	Subflooring, roof and wall sheathing, underlayment, siding (with overlay)	Underlayment, cabinets, furniture, stair treads	Cabinets, furniture moldings	Siding, paneling, underlayment
Wood parts used	Wafers or flakes from whole trees	Particles from "waste" wood	Wood fibers	Wood fibers
Resin	Phenol-formaldehyde (some isocyanates)	Urea-formaldehyde[1]	Urea-formaldehyde	Phenol-formaldehyde
Grades	OSB is performance rated by APA: sheathing, siding, Sturd-I-Floor, structural 1 (for shear walls)	High, low and medium density. Special underlayment grade	Not applicable	Standard, tempered and service. Also available smooth one side (s1S) or two sides (s2S)
Density (PCF)	40	High: over 50 Medium: 40 - 50 Low: below 40	40 - 50	60 - 65
Approx. fastener holding (face)	20 lb. for 6d nail[3] 15 lb. for 8d nail[4]	225[2] lb. for #10 screw	355 lb. for #10 screw	150 lb. for nail-head pull-through of a 6d nail[5]
Approx. fastener holding (edge)	Edge-nailing not recommended	200[2] lb. for #10 screw	290 lb. for #10 screw	Not recommended
Maximum thickness swell (24-hour soak test)	6%	Not available	3.7%	10%

1. A phenol-bonded particleboard is available but rare 2. Medium density particleboard 3. Through ½-in. thickness 4. Over ½-in. thickness 5. 3/32-in. wire diameter with 17/64-in. head diameter

Photo this page: Susan Kahn

gets into everything," says Frane. He says that people who use it regularly worry a lot about the dust. Even if they wear dust masks, "they're blowing their noses for three days afterwards." For this reason, it's a good idea to wear a respirator when machining MDF. One carpenter I met made a point of wearing coveralls when using MDF. He didn't want to bring the dust home where his kids could breathe it.

Hardboard—Hardboard is the oldest of the composition panels. It was first produced in the early 1920s by W. H. Mason, founder of Masonite Corp. It's available in three grades: standard, tempered (chemically and heat treated to improve stiffness and hardness) and service (not as smooth or strong as standard but lighter). It can be manufactured with one or two sides smooth (S1S OR S2S). Like MDF, hardboard is made from wood fibers rather than discrete particles. The main differences are that hardboard is denser and that it's bonded with a phenolic resin. According to Louis Wagner, technical services director at the American Hardboard Association (AHA), 60% of all hardboard in this country is manufactured as siding. The other 40% goes into everything from door skins to furniture to auto interiors.

Sheet hardboard comes in ¼-in. thicknesses. Most builders are familiar with it as an underlayment for resilient flooring, but because particleboard is more stable, more available and less expensive than in the past, hardboard's share of the underlayment market has all but evaporated. According to Wagner, hardboard underlayment isn't easy to find. Most manufacturers, in fact, don't even bother making it (though it's worth noting that Masonite makes an underlayment grade hardboard in 4x4 sheets).

Aside from underlayment, most builders I've talked with are at a loss for what to do with this product. About the most common use for sheet hardboard I found was among remodelers as protection for hardwood floors during construction. One builder told me that the ¼-in. thickness made it too floppy for anything else and that anything that needs to be ¼ in. might as well be plywood. But floppiness can be an advantage in some situations. S1S hardboard has been used for curved concrete forms where a smooth finished surface is needed. This application requires that the inside of the form be coated with an oil release agent. And one homeowner I talked with said he used S1S hardboard to line his laundry chute.

When installing any hardboard product, the AHA recommends that you use nails with flat, full round heads and that you don't overdrive them. Because hardboard is so dense, overdriving a nail will cause the surface to lump up around it. The AHA recommends not using finish nails, even on hardboard siding. They're designed to fasten materials that are thicker than hardboard siding's 7/16 or ½ in.

One area where the product may be underutilized is as a web material for I-beams, an application that's much more common in Europe than in the U. S. In tests conducted by W. C. Lee, a professor in the forestry department at Clemson University in South Carolina, hardboard was found to have a higher interlaminar shear strength than either OSB or plywood. Interlaminar shear strength measures the resistance of a material's layers to being pushed apart (as opposed to the more familiar edgewise shear, which measures a material's resistance to racking). It's the type of stress most likely to occur in a composite I-beam. □

Charles Wardell is assistant editor of Fine Homebuilding.

Particleboard and formaldehyde

Formaldehyde is everywhere: it puts the permanent in permanent press, the absorbency into paper towels and stretches the shelf life of certain paints and coatings. American industry uses about a billion pounds of the stuff every year, though most household products contain only trace levels. The biggest source of indoor formaldehyde by far is wood products containing urea-formaldehyde resins.

Airborne formaldehyde at sufficient concentrations can cause watery, burning eyes and make breathing difficult. Long-term exposure can permanently sensitize a person to a host of other everyday chemicals. High concentrations can cause asthma attacks in people with asthma. And formaldehyde is a suspected human carcinogen. The HUD standard for manufactured housing limits indoor formaldehyde concentrations to 0.3 parts per million (ppm), while the American Society of Heating, Refrigeration and Air Conditioning Engineers (ASHRAE) recommends a maximum exposure of 0.1 ppm. Irritation has been observed at levels as low as 0.01 ppm. The point is that there's a great variation in individuals' sensitivity levels.

Uncured urea-formaldehyde resin consists of short chains of urea and formaldehyde molecules, along with some free formaldehyde. Under the heat and pressure of the press, the free formaldehyde links these short chains into longer ones. Afterward, some free formaldehyde that remains in the pores of the board can migrate to the board's surface and then to the surrounding air. How much free formaldehyde remains in the board depends on how much was in the resin. Today's resins contain less free formaldehyde than those from 10 years ago, so they emit less to the surrounding air.

The wood products industry has voluntarily adopted the HUD standard as an emission standard for urea-bonded panels. The standard sets a limit of 0.3 parts of formaldehyde per million parts of air in a room containing 0.13 sf of particleboard per cu. ft. of room space—which about describes the amount of underlayment in a typical room with 8-ft. ceilings. Tests are conducted at 50% relative humidity, 77° F and with a ventilation rate of 0.5 air changes per hour. Homes with more particleboard—or those that are tighter or more humid—could have higher levels of offgassing.

Not everyone is happy with the HUD standards. Thad Godish, director of the Indoor Air Quality Research Lab at Ball State University in Muncie, Indiana, told me that he would still not recommend particleboard "under any circumstances." Godish claims that the HUD standard is based not on health considerations, but on what the manufacturers can achieve at a relatively low cost. He also said that though manufacturers cover their particleboard cabinet carcases with laminates, the finish used on almost all wood-faced cabinets has formaldehyde in it.

You can sidestep the formaldehyde question altogether by using exterior-grade panels made with phenol-formaldehyde resins; offgassing from these resins is negligible. Phenol-bonded products include waferboard, OSB and exterior-grade plywood. At least one manufacturer, Medite Corp. (P. O. Box 4040, Medford, Ore. 97501; 503-773-2522) makes a formaldehyde-free MDF. It's bonded with a poly-urea matrix. —*C. W.*

For more information

The following trade associations distribute information about the products included in this article, including technical data sheets describing product and performance standards and proper means of installation. Most of their publications are available for free or for a nominal fee.

OSB:

American Plywood Association
P. O. Box 11700
Tacoma, Wash. 98411
206-565-6600

Structural Board Association
(formerly The Waferboard Association)
45 Sheppard Ave. E., Suite 442
Willowdale, Ontario M2N 5M6
416-730-9090

Particleboard and MDF:

National Particleboard Association
18928 Premiere Ct.
Gaithersburg, Md. 20879
301-670-0604

Hardboard:

American Hardboard Association
520 North Hicks Rd.
Palatine, Il. 60067
708-934-8800

Glue-Laminated Timbers

Designing and building with engineered beams

by Stephen Smulski

Until recently, glue-laminated timbers, or glulams, were used chiefly for commercial and institutional structures. Most people know them as the exposed framework of shopping centers and churches. Home builders generally used steel for long spans and sawn timbers for exposed frames.

But times are changing. Open floor plans remain popular with homeowners, as do exposed beams with long, clear spans. Sawn beams that fit the bill are usually milled from large, old-growth trees, which are becoming scarce and expensive. As a result, more builders are turning to engineered-wood products, including glulams (for a comparison of glulams and competing structural beams, see the table on p. 105). In residential construction, glulams can serve as strong and attractive exposed structural members with long, clear spans. Their uses include arches, headers, girders, ridge beams, joists and rafters (photo above). Not surprisingly, glulam sales are expected to double over the next ten years.

Longer, wider, deeper—Glulams are made in a wide range of sizes, stiffnesses and strengths by face-laminating dimension lumber with structural adhesives. The lamination process yields beams that are longer, wider and deeper than sawn beams, so a glulam uses wood more efficiently than a sawn beam. The wood in a glulam is harvested from second- and third-generation trees, taking some of the pressure off old-growth forests.

The laminations in a glulam beam consist of 6-ft. to 20-ft. lengths of 2x dimension lumber (1x for tightly radiused arches), kiln-dried to an average 12% moisture content. The lumber is end-joined with structural finger joints and a waterproof, synthetic-resin adhesive (melamine, phenol, resorcinol, or phenol-resorcinol-formaldehyde). In the U. S., most glulams are made from Southern yellow pine or Douglas fir, although redwood, Alaska yellow cedar and Western red cedar are also used. In theory, there's no limit to a glulam's size, but like most building materials, they're shipped by truck and rail. This limits lengths to 100 ft. or so (oversized trailers are used to ship glulams longer than 60 ft. over the road).

Almost all glulams are horizontally laminated. Adhesive is spread on one wide face of each lamination; the laminations are then placed in a clamping jig. The adhesive cures at room temperature and develops sufficient strength so that clamps can be removed in 6 hours to 24 hours. The glulam is then set aside for a brief conditioning period during which the adhesive develops full strength.

After conditioning, the glulam is surface-planed and crosscut to length as needed. Tapered glulams and arches are machined to their final shape. Edges are eased, and splits, knots and other surface imperfections are plugged and smoothed to meet appearance-grade requirements. Some manufacturers may bore connector holes and apply end sealers, primers, or surface finishes. Some also supply fasteners, hangers and engineering assistance upon request. Before shipping, glulams are wrapped in kraft paper or a housewrap-type material that keeps the beam dry, yet lets it breathe. The wrapping also prevents surface marring during storage, transport and erection.

Stiffness and strength—Because the adhesive used in a glulam is stronger than the wood itself, the beam's individual laminations act

From *Fine Homebuilding* magazine (December 1991) 71:55-59

together as a single unit. Allowable design values for glulams are higher than those for sawn timbers of the same species, depth and board footage (large photo, facing page). This superior performance is engineered into glulams during fabrication in subtle, but effective ways.

Straight glulams, for example, are made with a measured camber, or crown. Beams are installed "crown up" and the crowned face is always clearly marked "TOP." Because the camber is equal to the amount the beam will deflect downward under the structure's dead load, the beam will ultimately lie flat or retain a minimal residual camber.

A glulam's performance is further enhanced by being made from dry wood. Dry wood has higher design values than green wood, but large sawn timbers usually dry in place, where they warp, check, split and release water vapor. This can weaken the timber, raise energy costs and encourage condensation problems in the house during the first heating season. The checking and splitting can be minimized by slowly kiln-drying 6x6 and larger timbers under carefully controlled conditions, but the process is too expensive (70% of the energy that goes into making a piece of dimension lumber is used up in the kiln; the larger the pieces, the more energy the drying process consumes). Because a glulam is made with dry lumber, its potential for warping, checking and moisture release is minimal.

The most crucial part of the engineering process, however, is the grading and sorting of the lumber that will go into a beam. Glulams are engineered like I-beams, which means that strength is built in where it's needed most. In a straight, simply-supported beam, for example, the bottom surface will be stressed in tension and the top surface in compression. Consequently, glulam manufacturers place the highest-quality lumber in the bottom lamination and the next highest-quality in the top (left photo, facing page). Lower-quality lumber goes into the beam's interior. This lets the manufacturer restrict most stiffness- and strength-reducing defects (such as knots and slope of grain) to the least critical parts of the beam.

But while this placement method works for a simply-supported beam, there are situations where it's inappropriate. In a cantilevered or multiple-span beam, for example, both surfaces will be in tension at some point—the bottom will be in tension between the supports, and the top in tension over the supports. So cantilevered and multiple-span glulams must be made with what's known as a balanced layup. These beams use lumber of equivalent design strengths for their top and bottom laminations. (Tension ratings for a stock beam are usually 2,400 psi on the bottom and 1,200 psi on the top; a balanced beam would be rated for 2,400 psi on the top and bottom.) Balanced layups are usually special-order items.

Is all this lumber sorting and placement worth it? You bet. Glulam members with strategically positioned lumber are 10% to 20% stiffer, and up to 100% stronger in bending than those with randomly located lumber.

Sizes and grades—Glulams are produced in three appearance grades—Industrial, Architectural and Premium. Allowable design values for all three grades are identical, but the higher grades have fewer surface defects (knots, checks, splits, wane, planer skips and saw marks).

Preservative-treated glulams are available as a special-order item. They should be used for all exterior applications and for interiors with consistently high relative humidities, such as pool enclosures. When oil-borne preservatives such as creosote or pentachlorophenol are used, the glulam is treated after laminating. With water-borne preservatives like CCA, the individual laminations are treated first, then redried to a 12% moisture content and glued. Otherwise, severe checking would occur as completed members dried after treatment.

Manufacturers can also make glulams with a 1-hour fire-resistance rating. This is done by removing one core lamination from the middle of the beam and adding an extra tension lamination at the bottom. During a fire the extra tension lamination becomes the sacrificial lam. If it's damaged, there will still be other tension laminations to back it up. Even stock glulams have excellent fire-resistance ratings, however; in a fire a glulam will retain its strength significantly longer than a comparable length of exposed steel.

How to buy them—To order a glulam, start with your local building-materials supplier. You may also be able to order what you need directly from a glulam manufacturer (for sources of design information, see the end of this article). You'll have to specify width, depth, length, camber, appearance grade, allowable stress rating and the desired finish. You'll also want to identify any connectors that call for holes to be drilled in the glulam (more on that in a moment). If all that seems like a bit much, most fabricators will provide engineering assistance. Send them a copy of your plans (or give your plans to the retailer if you're working through one), and their engineering staff will figure out what you need.

When specifying glulams, pay close attention to their actual (as opposed to nominal) width and depth. Because glulams are edge-planed *after* lamination, the finished product isn't as wide as the original pieces. For example, a stock Douglas fir glulam that's made from 2x6 dimension lumber will finish out at 5⅛ in. rather than 5½ in. Southern yellow pine glulams will be a bit narrower still. A stock Southern pine glulam won't be as deep as its Douglas fir counterpart either, because highly resinous Southern yellow pine boards are surface-planed *before* laminating to present a fresh surface to the glue. Southern pine glulams are laid up with lumber 1⅜ in. thick, rather than 1½ in. Recognizing the confusion this can cause, some manufacturers now make products the size of Douglas fir beams by starting with oversized pieces of Southern yellow pine.

Placing a beam—Glulams are usually big, and big means heavy. American Institute of Timber Construction (AITC) and American

Sizing up a structural beam

by Christopher F. DeBlois

Most residential loads, even large ones, are small enough to be supported in a number of ways. For example, the engineering firm I work for recently acted as structural engineers for a development of narrow, one-story villas. The owner of one of these wanted to extend the main roofline to enclose a small indoor pool. We were asked

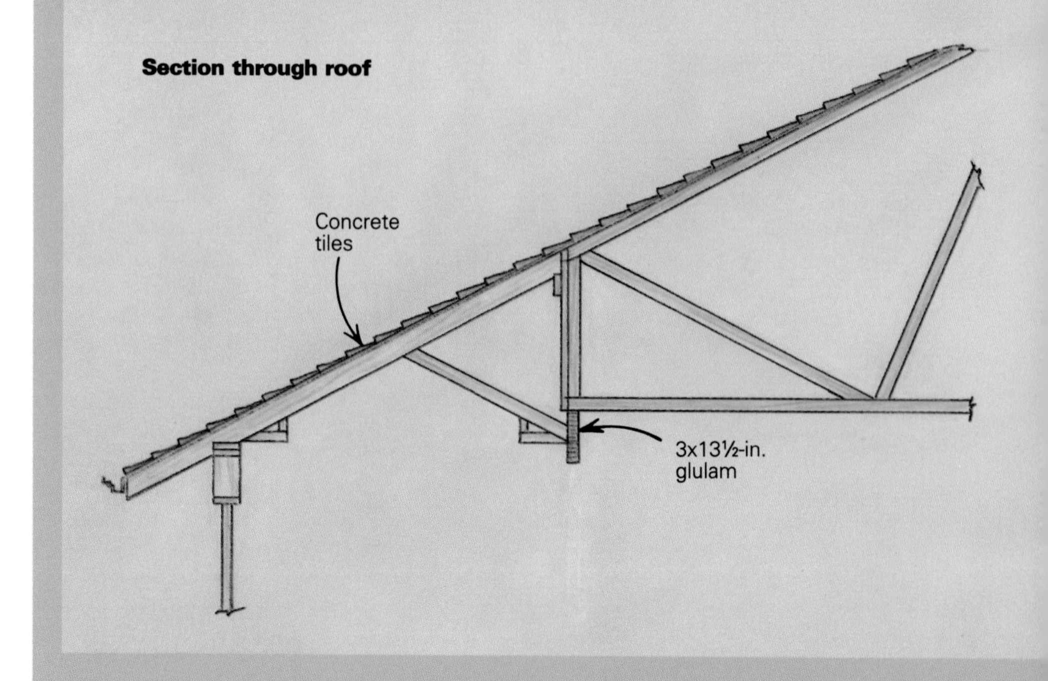

Drawings: Christopher Clapp

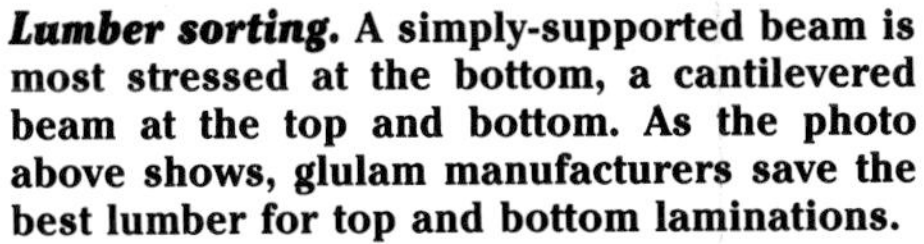

Lumber sorting. **A simply-supported beam is most stressed at the bottom, a cantilevered beam at the top and bottom. As the photo above shows, glulam manufacturers save the best lumber for top and bottom laminations.**

Heavy construction. **Glulams can shoulder heavier loads than timbers of comparable size. Otherwise, they're no different to build with.**

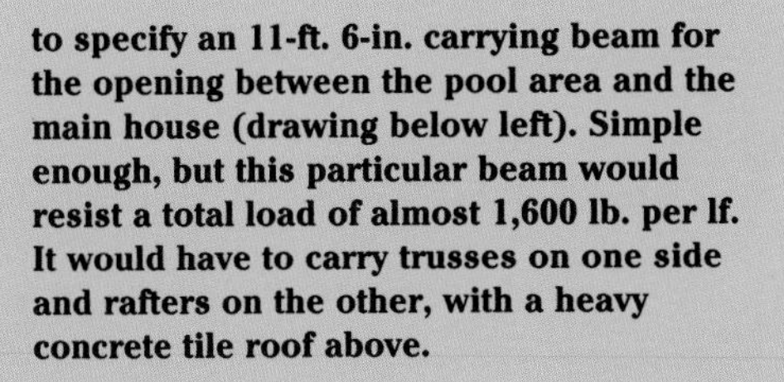

to specify an 11-ft. 6-in. carrying beam for the opening between the pool area and the main house (drawing below left). Simple enough, but this particular beam would resist a total load of almost 1,600 lb. per lf. It would have to carry trusses on one side and rafters on the other, with a heavy concrete tile roof above.

Although we ended up using a 3-in. wide by 13½-in. deep glulam, that certainly wasn't the only option. The table below summarizes the equivalent choices available to us. It should serve as a good approximation of the relative merits of competing beams.

When you evaluate beams for residential construction, it's important to compare apples with apples. I usually only consider beams 3½ in. wide or less—very often, there's just no room for anything wider (this case was an exception in that wider beams were acceptable). I also try not to specify anything that's too hard to obtain. For instance, I won't consider nail-laminated beams made with anything larger than 2x12s.

Prices in the table are manufacturers' retail prices in Atlanta. Prices and weights are for an 11-ft. 6-in. length, but some of the beams may not always be readily available in this length. To get an 11-ft. 6-in. steel beam, for instance, you might have to buy a twenty-footer. These prices don't include installation labor, which can be a big chunk of the final cost: heavier beams require more workers, and some beams are harder to detail than others. LVL and nail-laminated beams must be nailed, and a flitch beam bolted together on site. It takes more work to hang joists from a piece of steel than from a solid beam. And if the beam you choose is much deeper than it is wide, you'll have to brace it to ensure its lateral stability (see *FHB* #70, p. 14).

Type of Beam	Size (in.)	Weight (lb.)	$/lineal ft.	Total $
Solid timber	6½ × 14	228	10.50	126.00
Glue-laminated timber (glulam)	3⅛ × 13½	120	7.00	81.00
Parallel strand lumber (PSL)	3½ × 12	151	8.50	98.00
Laminated veneer lumber (LVL)	3½ × 11⅞	122	6.40	74.00
Flitch beam	3½ × 11¼	320	17.40	200.00
Nail-laminated beam	7½ × 11¼	250	5.95	72.00
Steel I-beam	4 × 8⅛	173	9.07	105.00

Christopher F. DeBlois is a structural engineer with Nielsen/Uzun Structural Engineers in Atlanta, Georgia.

Photo above: Charles Wardell

Wood Systems (AWS) span tables give glulams' weight per lineal foot, but you can also figure them mathematically: Southern yellow pine beams weigh in at 36 lb. per cu. ft., Douglas fir at 35 lb. pcf, hem-fir and redwood at 27 lb. pcf. Some glulams can be installed without special equipment—if you have enough help. The 5-in. by 13¾-in. by 17-ft. Southern yellow pine glulam in my house weighs about 300 lb. and three of us lifted it into place by hand. Some jobs, however, require a crane or hoist. In these cases the glulam should be lifted in a webbed-belt sling. To protect the finished edges, place wood blocks between the sling and the glulam. Fine-tune the beam's final position with a rubber mallet, or with a hammer and wood block; otherwise you risk marring its surface.

Drilling and notching—There's only one important rule to remember when drilling and notching a glulam: don't do either without first consulting the fabricator or a qualified engineer (that includes drilling holes for connector bolts). Ideally, holes for connector bolts should be part of the beam's design. The reason for this caution is the typically high loads carried by a glulam. While the repetitive design of conventional framing spreads loads across many members, permitting liberal drilling and notching, a single glulam may carry a significant structural load without backup. You just can't bore and cut these things like you would dimension lumber (drawings below). Because shear is greatest at the ends of the beam, holes closer to the ends may lower the beam's shear strength, while those nearer the top or bottom may reduce compression or tensile strength. A rule of thumb is that a simply-supported glulam beam can tolerate a 1½-in. dia. hole at mid-depth of the central one-third of its span.

Notching causes similar problems. Notching a glulam can compromise its strength more than common sense might indicate. When you take a 2-in. notch out of a 12-in. deep glulam, you've effectively created a 10-in. beam. The closer the notch is to the center of the span, the more it weakens the beam. A glulam can withstand limited end-notching to provide clearance or to bring its upper surface level with adjacent framing, but AITC restricts compression-side end notches to 40% of beam depth, and tension-side notches to 10% of depth. Don't even think about notching the tension side of a beam anywhere but at its ends. When making end notches, use tapered, rather than square cuts. The latter can concentrate stresses that may split the beam; the former will reduce that stress concentration. Even if you're within all these guidelines, you should still check with your fabricator or engineer before proceeding. Of course, the best practice is to design the structure so that you won't have to drill or notch the glulam at all.

End supports—The wood fibers at each end of a simply-supported glulam are subjected to compression stresses from imposed loads, so the bearing area has to be large enough to keep these stresses from crushing the wood. The bearing area provided by a 5½-in. wide top plate is generally adequate in residential construction, although it depends on the particular beam and loading situation. A wood or

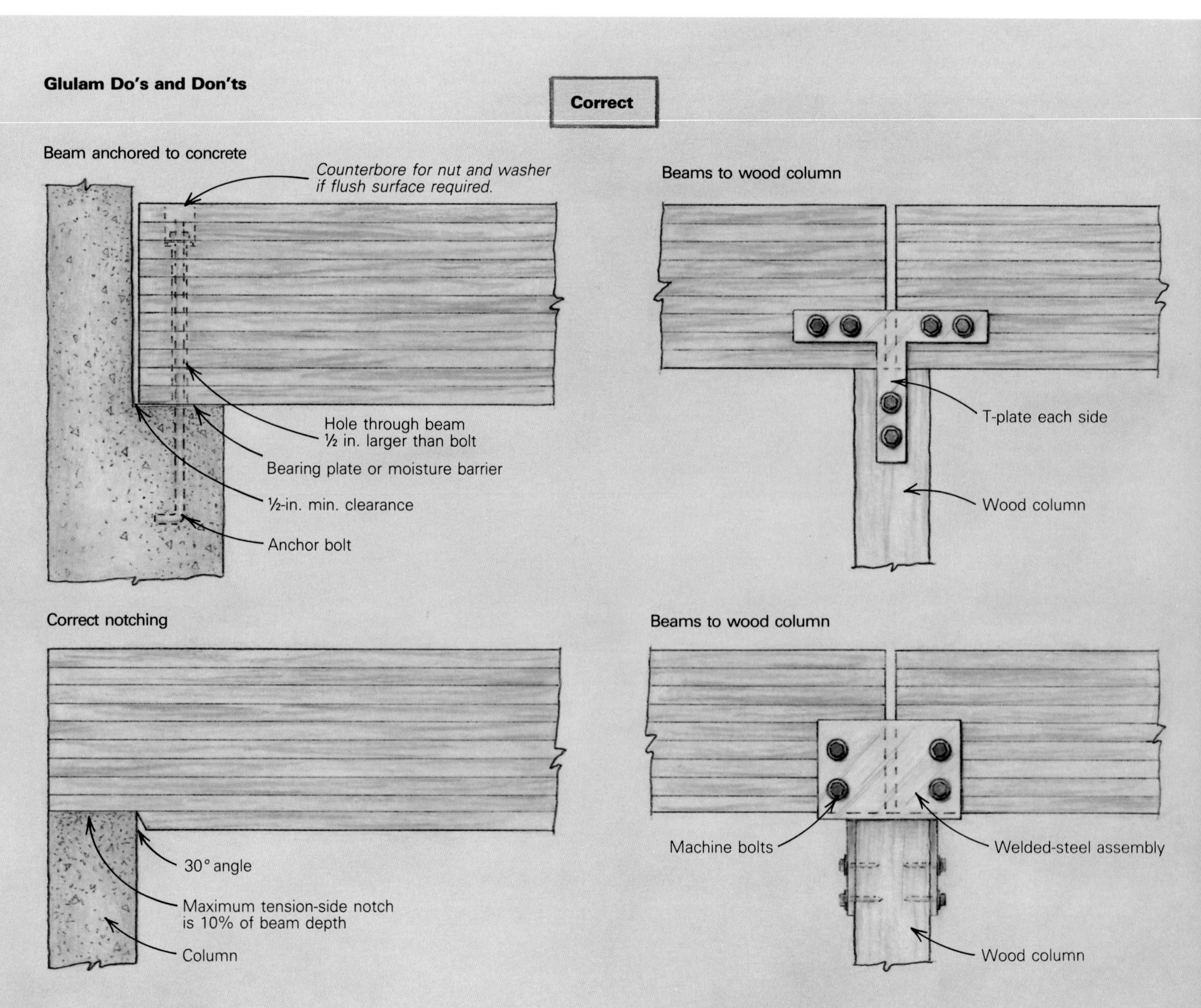

metal column whose cross section is at least equal to the bearing area should be installed under the top plate at the point of bearing and should transfer the load directly to the foundation. The ends of the glulam may need bracing or fastening to prevent rotation and uplift.

For glulam-to-glulam and framing-to-glulam connections, use standard or architectural heavy-gauge metal hangers. You should be able to order these through your glulam supplier. Fasten them with through-bolts, lag screws or spikes. When using spikes, drill pilot holes to eliminate splitting.

Glulams and moisture—When fastening a glulam, it's critical that the connection permit free swelling and shrinking of the member with seasonal variations in moisture content. Because of its large size, seemingly small variations in moisture will make a glulam swell and shrink quite a bit. During the summer, the seasonal rise in relative humidity in my New England home can cause my Southern yellow pine glulam to grow ⅛ in. in width and ¼ in. in depth. Under such conditions, the stress caused by misplaced fasteners could easily split the wood; if these splits are large or numerous, they could drive the beam to the brink of failure. It's also important not to put more fasteners into the beam than are needed. Coating the beam with a finish will retard moisture pickup and reduce dimensional changes.

On site, store glulams off the ground and cover them with a protective tarp. Create a tent by weighting the top of the tarp and letting its sides drape loosely. The idea is to promote air flow and to keep the tarp from trapping moisture. During long construction delays, end-coat the beam with a couple of coats of paint or water-repellent. To guard against surface marring, keep the fabricator's protective wrap on until after the beam has been installed and the drywall taping and interior painting is done.

For more information—Virtually all glulams made in the U. S. conform to the manufacturing and quality-control specifications of the American National Standards Institute (ANSI A190.1-1983). Two of the largest independent agencies that provide inspection and testing services to glulam manufacturers are the American Institute of Timber Construction (11818 SE Mill Plain Blvd., Vancouver, Wash. 98684; 206-254-9132) and American Wood Systems (P. O. Box 11700, Tacoma Wash. 98411; 206-565-6600), a related corporation of the American Plywood Association. An AITC or APA-EWS quality mark stamped on a glulam signifies conformance with ANSI quality standards.

The classic reference for glulam and heavy timber construction is AITC's *Timber Construction Manual* (J. Wiley & Sons, Inc., 1 Wiley Dr., Somerset, N. J. 08875; 800-225-5945, $59.95 plus shipping). Two free publications you'll also want for your bookshelf are AITC's *Glued Laminated Timbers for Residential and Light Commercial Construction*, and AWS's *Product Application Guide—Glulams.* □

Stephen Smulski is an assistant professor of wood science and technology at the University of Massachusetts at Amherst. Photos courtesy of the American Plywood Association, except where noted.

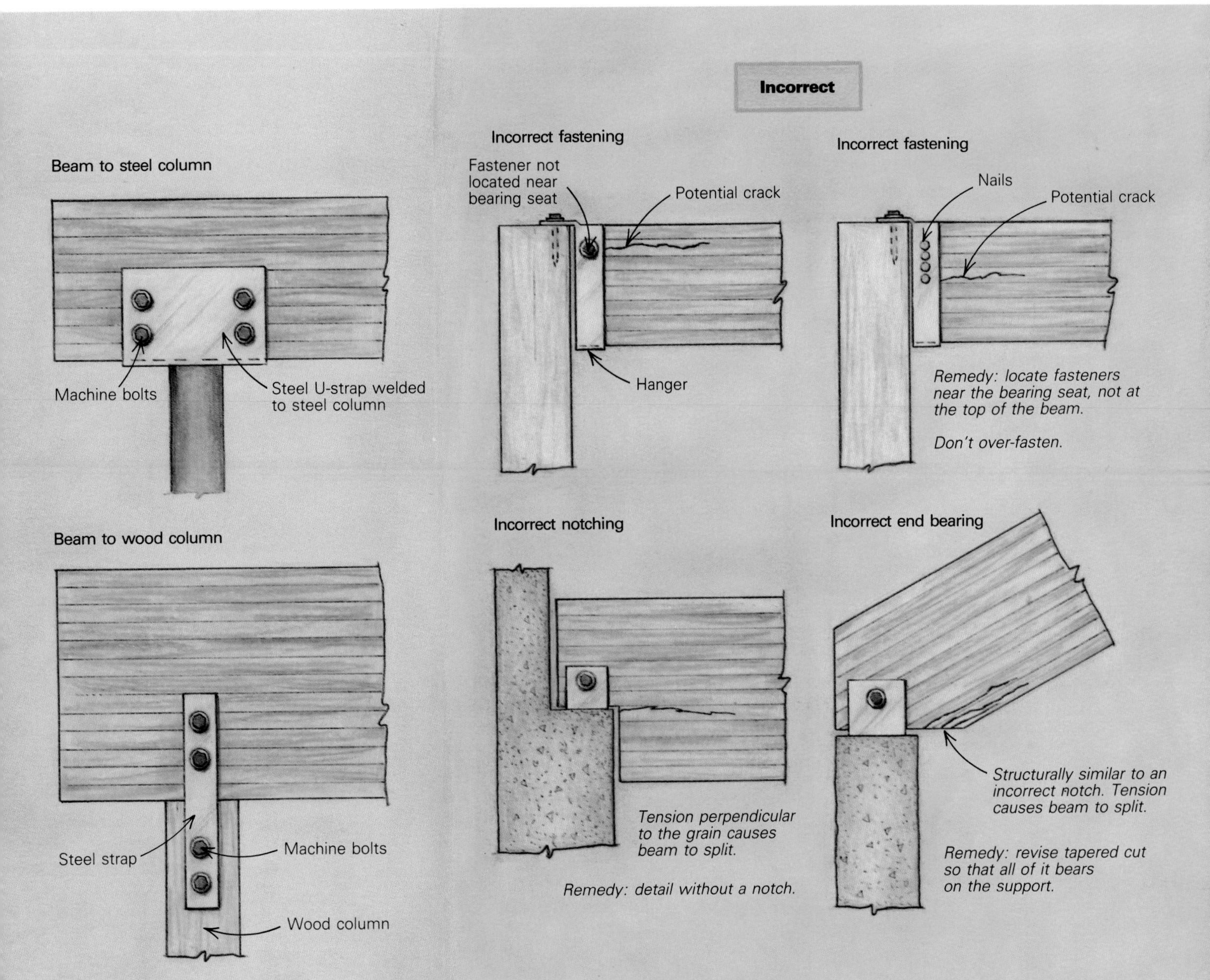

Drawings above rendered by Chris Clapp after originals from the American Institute of Timber Construction

Laminated-Veneer Lumber

Solid wood gets the short end of the stick

by Mark Feirer

Will you still be building houses in 2053? Only then will you be able to use lumber from Douglas-fir seedlings planted today. That simple fact, combined with the construction industry's appetite for lumber, has prompted considerable research into ways of making better use of the trees we have. The most successful methods literally take trees apart and glue them back together again in more efficient forms. These days it's hard to imagine a house that doesn't use plywood (which is simply sheets of veneer glued together) or particleboard (wood chips or flakes glued together).

Those products have revolutionized the way houses are built, and changes of a similar magnitude may be here. Manufacturing expertise and adhesive technology have brought us a new generation of wood products that promises to push solid lumber a little further off the job site. If you haven't used them yet, you probably will soon. Laminated-veneer lumber products are here (photo above).

Photo: Courtesy Trus Joist

The wood I-beam is part of a new generation of structural building materials that rely on parallel-laminated wood veneer and high-strength adhesive to make them light and stiff. Wood I-beams are increasingly being used instead of solid-wood rafters and joists. The chords of the I-beam rafter shown above are made from parallel laminations of veneer, and the web is cross-laminated structural plywood.

Good-bye 2x?—Laminated-veneer lumber (LVL) is a layered sandwich of wood and glue. Once it is fabricated into billets of various thicknesses, it can be cut at the factory into stock for headers and beams or into flanges for wood I-beams. Veneer lumber may look like plywood in cross section, but structurally, it's a very different product. Plywood is cross-laminated, meaning that the grain of each veneer layer runs at 90° to adjacent layers. This enables the panel to be thin and strong across both width and length, but also means that on every edge there's a good bit of exposed end grain. In laminated-veneer lumber, the grain of each layer of veneer runs in the same direction. This is called parallel-lamination, and it produces a material with greater uniformity and predictability than the same dimension material made by cross-lamination. First used during World War II to make airplane propellers, laminated-veneer lumber has lately become more familiar on the job site. Several companies now make LVL products (see sidebar, p. 110).

Just as solid 1x sheathing has been replaced by laminated-veneer sheathing (plywood), structural lumber may now be in the midst of being supplanted by laminated-veneer lumber. If that's the case, builders had better learn about these products: plywood didn't take long to become the dominant sheathing material in North America.

Promise and peril—Headers, joists and beams made in whole or in part from laminated veneer are becoming well established in the marketplace. The products have been thoroughly tested and are approved for use by the major code agencies in the U. S. and Canada. Some manufacturers tout the cost advantages of laminated-veneer products, particularly wood I-beams. But actually wood I-beams are a little more expensive than solid lumber. The savings, if any, comes during installation because less time is spent handling the material and drilling through it to run water and electrical lines. That's a level of savings most likely to fall to builders who specialize in speed and quantity. One price advantage custom builders will see, however, is one relating to length. Whether you buy a 6-ft. length or a 60-ft. length, the price per lineal foot often remains the same.

Laminated-veneer lumber products are considerably stronger when loaded parallel to the grain than an equal thickness of solid-wood lumber. You should consult an engineer, however, if you expect significant lateral loads. This is particularly important when using wood I-beams because they don't have the lateral stiffness that solid lumber does.

A major advantage for the small custom builder comes in the uniformity of veneer lumber. Because the products are composed of layers of veneer, defects in the wood are dispersed—you may see a minor defect here and there, but you'll never have to reject a length of laminated-veneer lumber because it has a loose knot running clear through it. Each piece of laminated-veneer lumber, whether an I-beam or a header, has been engineered to meet particular design criteria. The closest thing to this in solid wood is machine stress-graded (MSR) lumber. Because the manufacturing process is so closely controlled, the performance of LVL products is predictable.

There are other advantages, too, depending on how and what you build. Laminated-veneer lumber is typically available in lengths of up to 60 ft. and can be made in 80-ft. lengths if needed. That allows joists to reach from one side of a foundation to another without overlaps. The pieces are relatively light in weight, so one person can handle a length that would be a backbreaker if it were solid lumber. The downside, however, is that the light weight and flanged shape of an I-beam makes it an effective wind-catcher. Framing crews sometimes learn the hard way to take care when

From *Fine Homebuilding* magazine (December 1988) 50:40-45

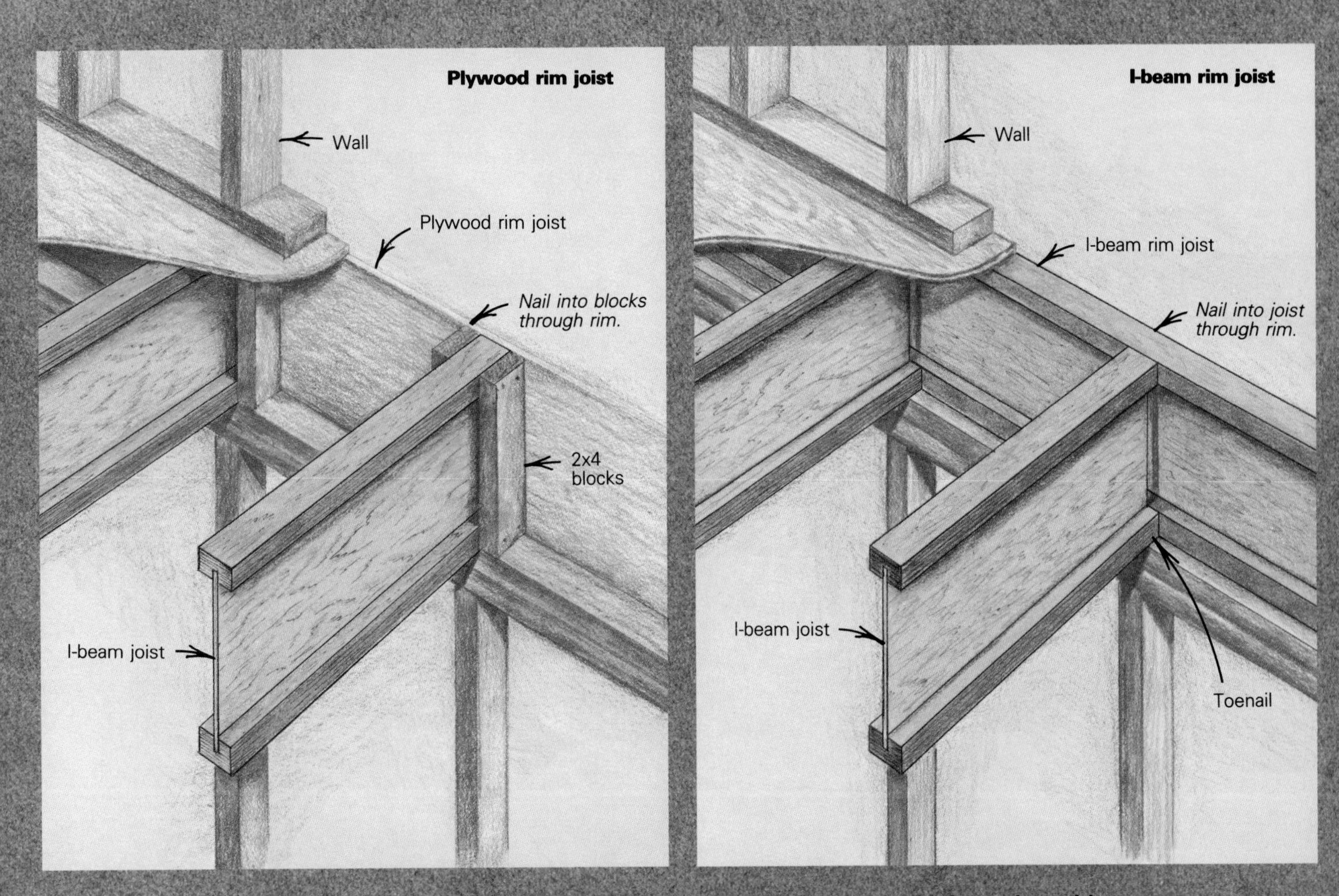

Engineered wood. ***Even the manufacturers admit that laminated-veneer lumber and wood I-beams aren't perfect substitutes for solid lumber, but it's clear that both products can be used to solve a variety of structural problems. Shown on this page are some common installation details for wood I-beams, perhaps the best known of this new generation of engineered-wood products. At the end of the article you'll find additional details for I-beams and laminated-veneer lumber.***

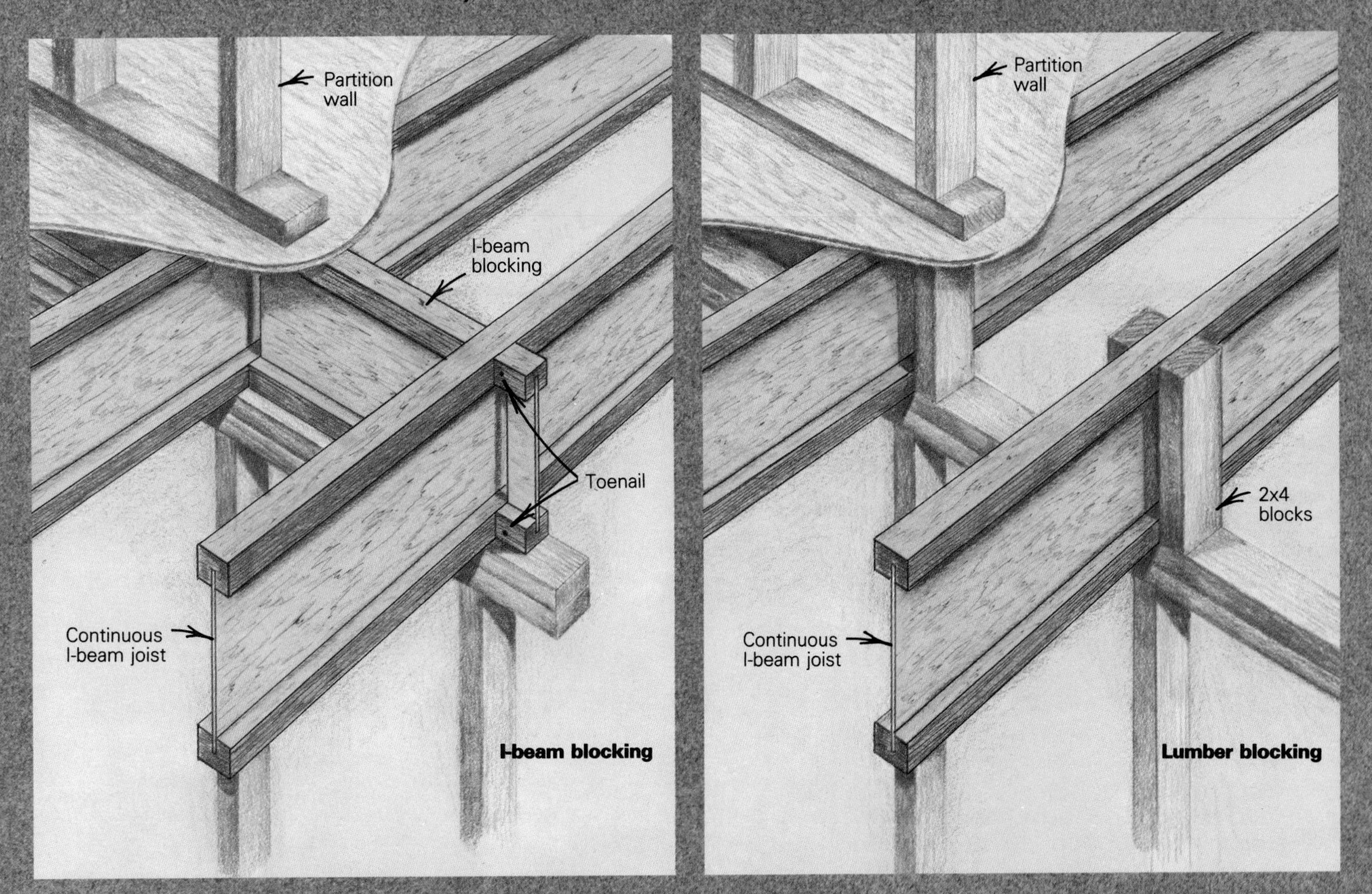

Sources of supply

Alpine Engineered Products, Inc.
P. O. Box 2225
Pompano Beach, Fla. 33061
Alpine makes residential I-beams, though they are not strictly a parallel-laminated product. Instead, they feature a plywood web and solid-wood chords.

MacMillan Bloedel Limited
1075 W. Georgia
Vancouver, B. C. V6V 3R9
MacMillan makes several parallel-laminated products, though they're a bit different from those of other companies. The basic material is called Parallam PSL, and it comes as beams, columns and headers. To make it, sheets of Douglas-fir veneer are sliced into thin strands of up to 8 ft. long, coated with adhesive and pressed into shape. The adhesive is then cured with microwave energy. Columns are available in sizes from 3½ in. square to 7 in. square. Beams and headers are available in sizes from 1¾ in. thick to 7 in. thick, and from 7¼ in. deep to 18 in. deep.

Mitek Wood Products
730 N.W. 107 Ave., Suite 400
Miami, Fla. 33172
Mitek makes residential I-beams with parallel-laminated chords and an oriented strandboard (OSB) web. The web is Weyerhauser's Structurwood.

Simpson Strong-Tie Co., Inc.
1450 Doolittle Dr.
P. O. Box 1568
San Leandro, Calif. 94577
Simpson makes a number of metal hangers for wood I-beams, including some for tying I-beams directly to masonry walls.

TECO Products Co.
12401 Middlebrook Rd.
Germantown, Md. 20874
TECO makes a full line of connectors and hangers for wood I-beams.

Tecton Laminates Corp.
709 N. W. Wall
Suite 104
Bend, Ore. 97701
Tecton's LVL beams and headers are made from scarf-jointed Douglas-fir veneers. The products range in thickness from ¾ in. to 2 in. and in length up to 50 ft.

Trus Joist Corporation
9777 W. Chinden Blvd.
P. O. Box 60
Boise, Ida. 83707
Trus Joist makes residential I-beams in 9½-in. and 11⅞-in. widths; Micro=Lam headers and beams in seven depths ranging from 5½ in. to 18 in. Headers and beams are 1¾ in. thick, and can be doubled or tripled on the job site as needed. Trus Joist also makes an I-beam for flat roofs that has a sloped top chord.

United Steel Products Co.
703 Rogers Dr.
Montgomery, Minn. 56069
The company carries a full line of connectors and hangers and will custom-fabricate for special applications.

working with the products on roofs and second-floor decks.

Compared to solid lumber, wood I-beams are relatively easy to run plumbing and electrical through. Prescored knockout holes are located every 12 in. along the web, and it takes just a hammer blow to create a uniform 1½-in. dia. hole for wiring and water lines (photo, p. 108). Larger holes can be cut with a saw. And forget about sighting down the length of a wood I-beam to spot a crown; I-beams and laminated headers are dead straight in any length. You can, however, order them with a camber.

Is laminated-veneer lumber perfect? Not quite. Like solid lumber, laminated-veneer lumber shrinks and swells. This isn't much of a problem when it comes to products such as wood I-beams that use LVL only for flanges. But the "solid" LVL products commonly used for headers have surprised some builders who expected complete stability. An 18-in. wide piece of LVL header stock, for example, is typically 17⅞ in. wide when it leaves the mill at 8% moisture content. But it can easily swell as much as ¼ in. (and sometimes ½ in.) by the time it's ready for use at the job site, making it ⅛ in. over its nominal dimension. That sounds familiar to users of solid-wood stock, but remember that whatever an engineered product will do, it will do uniformly. If one piece is wide they'll all be wide, so adjustments are easy to work around. Typically, most of the swelling in laminated-veneer lumber happens at the ends of the product because that's where end grain is exposed.

Another problem comes from the fact that laminated-veneer lumber products are relatively new on job sites, and subs don't always know how to work with them. Consequently, holes are cut for mechanicals in places that will weaken the structural integrity of a house. Large holes can be cut in a wood I-beam as long as they're cut in the right place; the manufacturer's literature gives details. Holes must *not* be cut in LVL headers, however.

Bonding the veneer—The ICBO (International Congress of Building Officials) approves any species of veneer for use in engineered wood as long as the composite meets basic requirements set up by the agency. Douglas fir is the most common veneer, though, and Southern yellow pine is also used.

The key element in the sandwich is the adhesive. To learn about it, I called Professor Terry Sellars of the Forest Products Utilization Laboratory at Mississippi State University. He was the organizer of last year's conference on engineered wood and specializes in the study of adhesives. The adhesive holding laminated-veneer lumber together is a phenol-formaldehyde resin that is, Sellars said, essentially indestructable. It's waterproof, as strong or stronger than wood, and laboratory and field research conducted over the years has proven its durability. Structural plywood uses the same phenol-formaldehyde, so it's already an old hand on the job site. Another glue commonly found in wood products is urea-formaldehyde, but high heat, such as that found in attics on summer days, degrades the glue, and exposure to water hydrolizes it. Urea-formaldehydes are also susceptible to fungus attack; this doesn't happen to phenols. In fact, the wood itself is likely to degrade well before the glue line does.

The making of an I-beam—To better understand how veneer and glue combine to make laminated-veneer lumber products, I visited a Trus Joist manufacturing plant in Eugene, Ore. The Trus Joist Corporation makes what is probably the best known of all the laminated-veneer lumber products. Called a TJI, it's an I-beam (the folks at Trus Joist prefer the term "I-joist") composed of laminated-veneer lumber flanges and a web of CCX or CDX structural plywood. Trus Joist also makes something called a Micro=Lam, which is a solid piece of laminated-veneer lumber, for use as header stock, floor beams and ridge beams. Based in Idaho, the company was the first to come up with a commercially viable product using parallel-laminated veneer. Right now, the Eugene plant is working 24 hours a day to keep up with demand.

The raw material for the TJI joist line and for Micro=Lams is ⅒-in. thick Douglas fir veneer, brought to the plant in a steady stream of flatbed trucks. By the time it arrives, the veneer has already been graded, and the 300-piece bundles (top left photo, facing page) are quickly forklifted into the plant. The manufacturing process is a mix of human and mechanical activity—machines checking the work of humans and humans returning the favor. I saw lots of quality checkpoints, and at any one of them the product can be rejected.

The first step is to dry the veneer to uniform moisture content. Individual sheets are hand-fed onto a three-level conveyor that runs continuously through huge drying ovens. It takes a tad under 13 minutes for a sheet of veneer to travel from one end of an oven to the other, and as each sheet exits, it is graded ultrasonically and checked for quality and uniformity of moisture content. An automatic paint-sprayer marks each piece—a dab of color to indicate grade, and a green stripe over the portions that are still a little too moist for specs. A person stationed at the end of the oven routes veneers into appropriate piles.

Sheets accumulate in bins by grade at the end of the dryer run and are forklifted over to the glue spreader. There each piece is fed by hand into an automatic glue-spreader, which coats the top of each sheet with a uniform layer of phenol-formaldehyde. The strongest, highest-grade veneers are directed to the top and bottom of the layup because these areas in the finished products will require the most strength (the upper and lower portions of the I-beam flanges). A person at the outfeed end wields a small roller on the end of a long handle, and touches up spots here and there that the machine missed. It's important to maintain uniform coverage and not use more glue than is absolutely needed.

The glued veneers feed onto a sliding table, which automatically layers them into a continuous multi-ply sandwich. The end of each ply overlaps the next one by about an inch. The sandwich is then fed into a machine where heat and pressure combine to cure the glue.

As the material exits the press, carbide-tipped circular-saw blades on each side trim the edges, and the resulting billet is cut to various lengths of up to 80 ft. At this point, the billets are directed either to the warehouse as Micro=Lams, or to the I-beam line where they receive additional attention.

Thirteen carbide-tipped blades are mounted on the gang saw, and this is the next stop for the billets destined to be TJI flanges. The saw takes a ¹⁄₁₀-in. kerf as it slices through the billet, and as individual chords exit the machine, they are visually inspected for uniformity and then fed into yet another machine. This one cuts a lengthwise groove in one face, eases the edges of the top corners and then, at high speed, spreads a thin layer of thermosetting pheno-resorcinol formaldehyde glue in the groove. In a complicated *pas de deux*, two continuous lengths of glued flanges come together around 4-ft. lengths of plywood-web material, which had previously been cut to width elsewhere in the factory. The end of each web piece receives a coat of glue, and ends are tightly butted together as the continuous chords wrap either edge. As the completed TJI exits the machine, utility knockouts are pressed into the web by a revolving wheel. Lengths of I-beams are then stacked in an oven to speed-set the glue, and from there, head into a warehouse where they are bundled, strapped with flat wire and stacked in the warehouse to await shipping (bottom left photo).

At the job site— Trus Joist typically ships 60-ft. lengths of I-beams to dealers. They're strapped in bundles wrapped in protective covering of reinforced plastic to keep them out of the weather. The dealer then cuts the product to approximate length for the contractor, trying to get the most efficient combination of cuts from each piece. It's the contractor who must trim the pieces to exact length at the job site. This may seem like wasted effort, but once laminated-veneer lumber leaves the perfect environment of the factory, it has to adapt to the imperfect world of the job site. A load of exactly cut I-beam floor joists delivered to a foundation that's out of square would cause expensive and time-consuming corrections.

Laminated-veneer lumber products take a little care at the job site. I asked Kevin Clemo, operations engineer for Trus Joist, about this. "Just keep them out of the weather as much as possible," he said, "and store them on edge." Like structural plywood, sunlight and a little rain won't hurt them but prolonged exposure could, particularly products without the strengthening shape of I-beams. Clemo suggested that the products be used within two or three weeks of arriving at the job site to ensure that swelling is kept to a minimum. Whenever the protective wrapping is missing, it should be replaced with water-resistant material that breathes; putting plastic over a bundle of wood can trap ground moisture inside the wrapping.

A wood I-beam is relatively weak in lateral strength, so storing one on its side could crack the glued butt joints that join individual lengths of the web. This could impair its strength, and the only way to repair it would be to nail/glue a plywood gusset over the affected joint.

Installation details—It's not tough to get used to working with laminated-veneer lumber. It accepts standard nails and cuts with standard tools. But because the sides of the flanges on I-beams are in a different plane than the web, the shoe of a circular saw can get hung up in the cut if you're not careful. Some builders get around this by slipping a scrap of wood between the flanges of the I-beam in order to even out its thickness. Others simply crosscut with a radial-arm saw.

Laminated-veneer lumber is simply wood and glue. In the photo at left, bundles of veneer await processing at a plant in Oregon. The numbers identify the grade of veneer and the mill supplying it. From this point on, the material can be tracked throughout the manufacturing process. This is part of the quality-control effort and allows engineers to correct any problems that might arise. Laminated-veneer lumber can be fabricated in lengths of up to 80 ft., and bundles 60 ft. long are typically shipped to wholesalers, where individual pieces are cut to lengths needed for specific projects. The I-beams in the photo below left are in the Trus Joist warehouse awaiting shipment. The pieces are nested together to save shipping costs. Though I-beams are most often used as joists, they can also be used as rafters (photo below).

Photo: Al Ebole

Nobody knows yet if there are any health hazards associated with the dust of cutting through lots of glue lines. But if you're going to be cutting a lot of the stuff, you might want to wear a dust mask.

The manufacturers of laminated-veneer products offer installation guides that show proper details of blocking, hanging and nailing the products (see sidebar, p. 110). But turn a creative builder and designer loose with any new material and you'll find all sorts of applications that don't show up in company literature. Builder Al Ebole and architect Daryl Rantis are just such people, and the custom house they're building west of Chicago is an encyclopedia of unusual techniques for working with I-beams and laminated-veneer lumber. A number of these, while approved by Trus Joist engineers, aren't yet in any catalog of details.

The laminated house—The house Rantis designed for the site in St. Charles, Il., uses Trus Joist headers and I-beams for everything from ridge beams and skylight headers to floor joists and garage-door headers. In fact, about the only place they're not used is in the walls—those are standard lumber studs. After numerous phone calls and a lot of discussion with Trus Joist engineers, Ebole came up with the details shown in the draw-

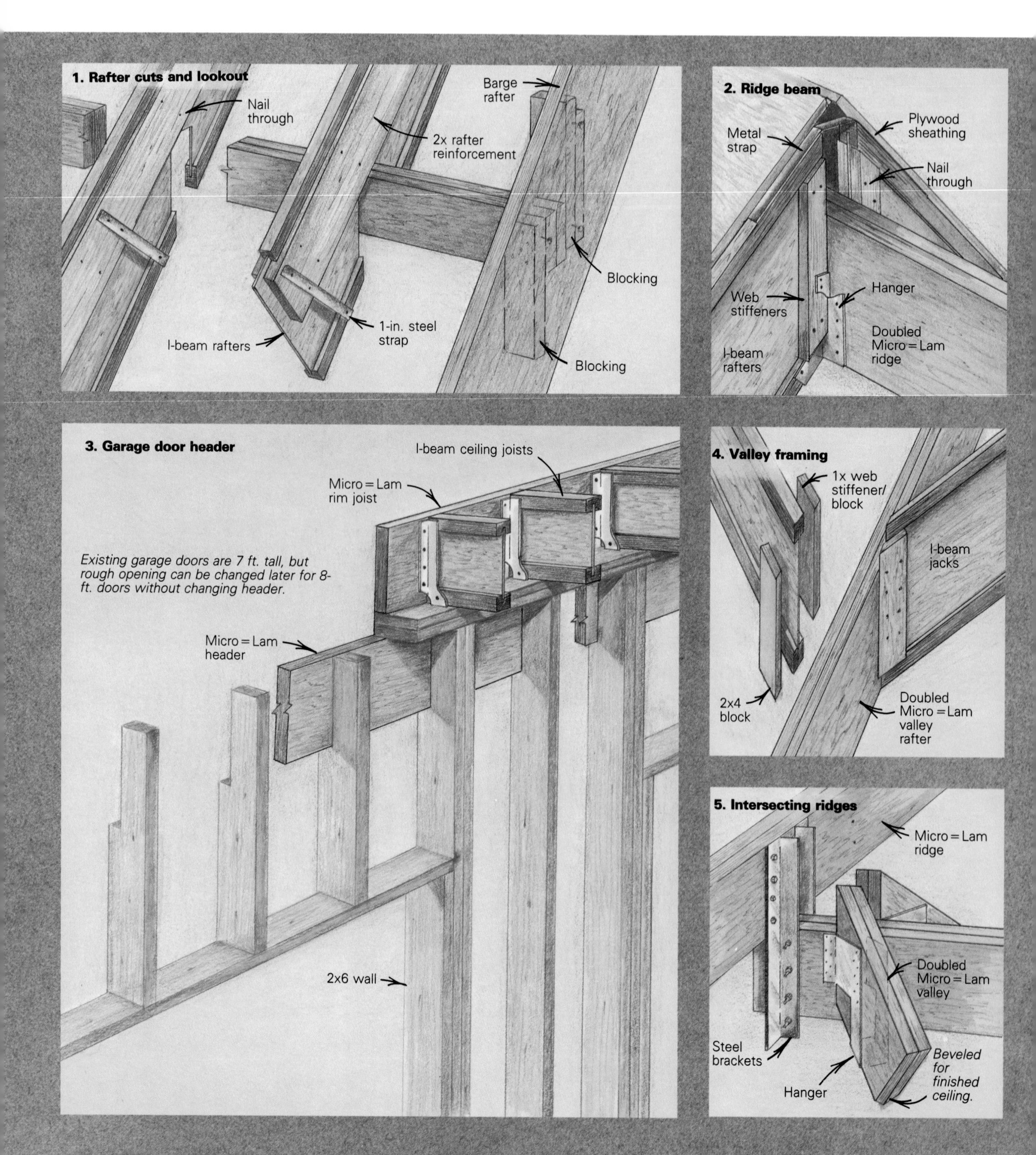

ings (facing page and below). Some of these details are unlikely to be called for on every project, but they show the versatility of engineered wood. Other details are bound to show up anytime LVL is used.

Ebole's crew used air guns on this project, and didn't have a lot of hand-nailing to do. But he told me that any kind of nail seems to hold better in laminated stock, and once you sink a 16d, you'll have a miserable time getting it out. As for cutting the products, Ebole has cut and ripped them just about every way imaginable.

The only real problem he had was with the Micro=Lams. The roof of the house is a fairly complex piece of work, and it took about six weeks to get it fully under cover. In the meantime, the 60-ft. long Micro=Lam ridge beam was exposed to alternate spells of rain and dry weather, with the result that it twisted and bowed. The same weather didn't seem to affect the TJI I-beams used as rafters, however, and Ebole now knows that the construction of these products makes them more stable than the Micro=Lams. Next time around, Ebole says he'd hire a laborer to paint any Micro=Lam stock brought to the job site. □

Mark Feirer is editor of Fine Homebuilding.

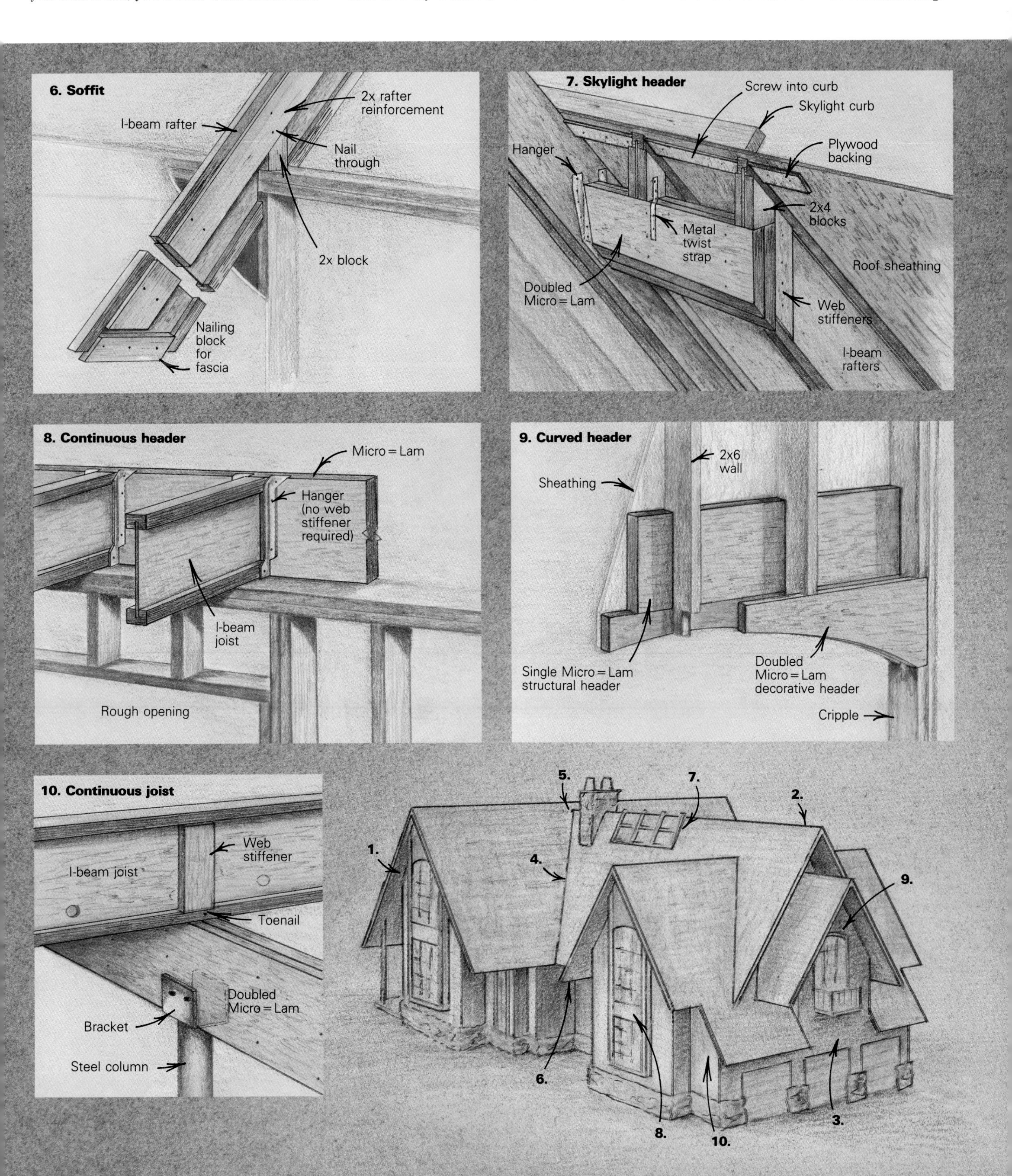

Preservative-Treated Wood

Lumber that can last a lifetime

by Stephen Smulski

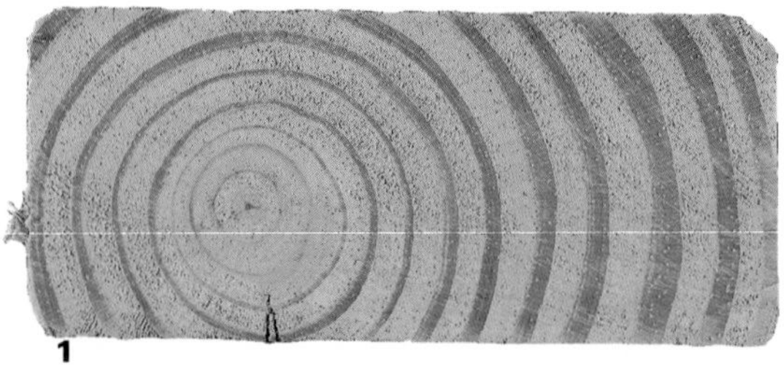

Cellulose and lignin, the stuff of which wood is made, are the two most abundant organic compounds on earth. Without the fungi and insects that biodegrade them, the landscape would be literally covered with downed and dead trees. The problem is that these relentless recyclers don't distinguish between wood lying on the forest floor and wood supporting your first floor.

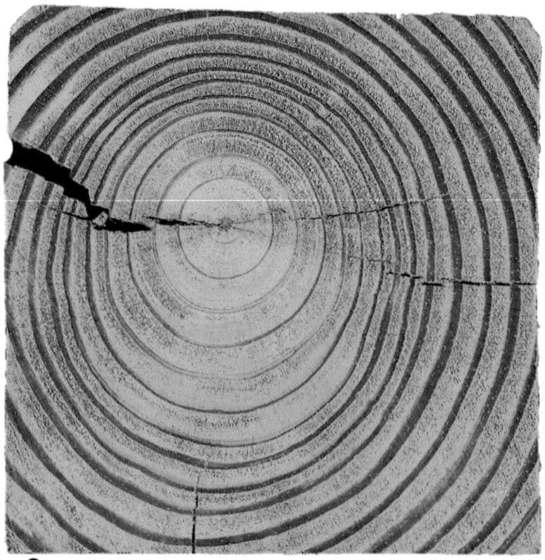

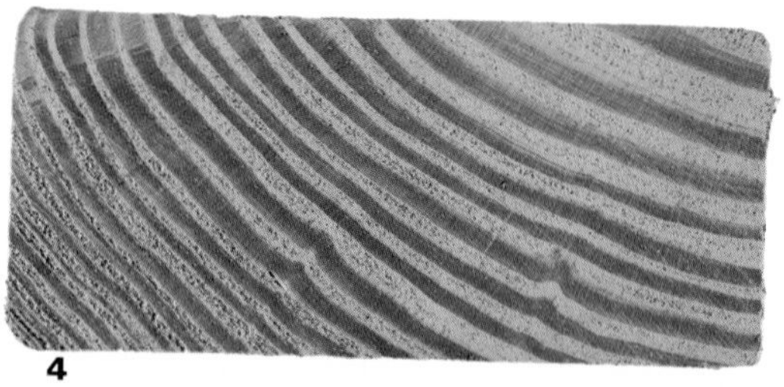

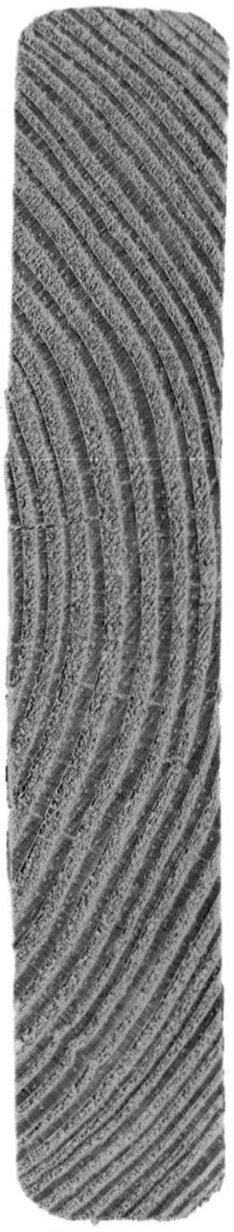

Man has been helping wood-destroying organisms make that distinction ever since the Egyptians first smeared wood funerary objects with cedar oil. But the wood-preserving industry didn't begin in earnest until the late 1800s, when America's railroads, faced with a shortage of naturally durable woods for crossties, started saturating lesser woods with creosote. Today, preservatives protect everything from poles, posts and piles to plywood, millwork and shingles.

Wood destroyers—Wood rots because it's being eaten by primitive plants called decay fungi, a process aptly termed the "slow fire." Other wood destroyers include carpenter ants, termites and dozens of beetles. Some gnaw tunnels in wood to create places to live, while others use it for food (for more on wood-destroying insects, see *FHB* #60, pp. 64-69). Crustaceans called marine borers also attack wood, burrowing into ships, piers and other wooden saltwater structures. Even bacteria can degrade wood under the right circumstances.

Preventing deterioration—Four conditions must exist before fungi or insects will attack wood: an oxygen supply; a continuous wood moisture content of at least 20%; a temperature in the 40° to 90° F range; and a food source, which is the wood itself. Eliminate any of these conditions and you eliminate the problem of decay indefinitely.

Of course it's hard to do much about temperature and oxygen, so the most effective and common "method" of preventing deterioration has always been to keep wood dry. This explains the long life of wood used indoors. In exterior use or other circumstances where wood can't be kept dry, the traditional method of delaying decay has been to use the heartwood of naturally rot-resistant woods, such as Western red cedar, redwood, baldcypress and white oak (see top chart, p. 117). Nature has kept these and other woods off the menu by

1

3

2

5

4

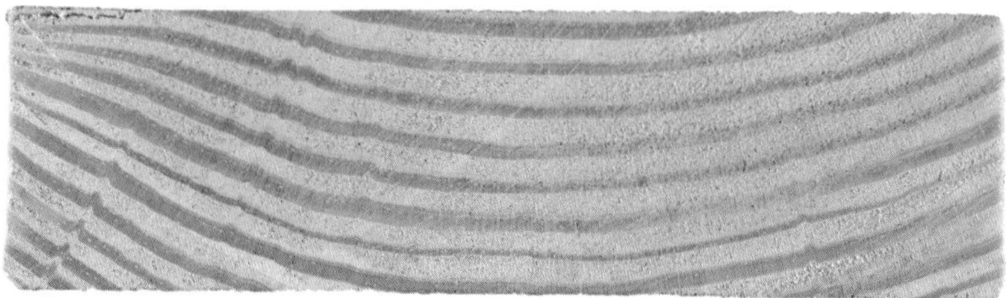

6

End grains. *Small stock is easily penetrated with preservative (photos 4 & 5). Industry standards require 90% penetration of the sapwood thickness, but as these photos illustrate, the preservative will not penetrate heartwood (photos 1, 2 & 3). Untreated zones retain the same color as the original board (photo 6).*

From *Fine Homebuilding* magazine (October 1990) 63:61-65

depositing in their heartwood unpalatable poisons called extractives. But supplies of naturally durable woods are too small to meet today's demand at an ecologically and economically acceptable price. In imitation of Nature's genius, the wood-preserving industry helps fill the demand by impregnating woods lacking decay-resistance with preservatives that can extend service life by 30 to 50 years or longer.

Both nonpressure and pressure processes are used to introduce preservatives into wood. The least effective are nonpressure processes: brushing, spraying and dipping. Brushing and spraying are usually limited to field treatment of wood during construction, or remedial treatment of wood in place. Millwork makers routinely dip window sash and other exterior trim in a water-repellent preservative. Poles, posts and piles are sometimes soaked for days or weeks, a process that is really nothing more than extended dipping.

Pressure treating—The amount of protection gained with these nonpressure methods is unpredictable. By far the most effective treatments are those where lumber is placed in a large pressure vessel and preservatives are forced into the wood cells (photos right). Though variations abound, the two basic pressure-treating processes are the *full-cell*, or *Bethell* process, and the *empty-cell*, or *Rueping/Lowry* process. With both, wood-cell walls are saturated with preservative. After treatment, cell cavities are filled with preservative in the full-cell process, but are nearly empty in the empty-cell process.

The full-cell process is needed to protect wood used in severe service environments. These include saltwater piles, utility poles and railroad ties.

In the full-cell process, a vacuum of 18-in. mercury is first drawn to remove air from wood's hollow cells. After the vessel is flooded with preservative, a pressure of 145 psi is applied, and the solution is driven deep into the wood. The pressure is later released and the excess liquid pumped to a holding tank. The treated wood is then removed for air drying.

The empty-cell process is used for most over-the-counter pressure-treated woods. In the empty-cell process, no initial vacuum is drawn. As the flooded vessel is pressurized, air inside the wood cells is compressed. When the pressure is released, the expanding air, aided by a small applied vacuum, kicks the preservative out of the cell cavities, leaving them nearly empty. Even though the cell cavities are empty, the cell walls are saturated with preservative.

Pressure-treating. **Lumber is placed in a sealed cylinder, and liquid preservative is forced into the wood cells under pressure. Arsenic-based compounds are among the most widely used preservatives. Wood treated with these chemicals has the familiar greenish-blue tint.**

Regardless of which process is used, green wood is generally dried to around 20% moisture content before treatment. Otherwise, the water that saturates its cells would inhibit absorption of preservative. Penetration depends on the type of wood and the size of the lumber being treated. Species that are difficult to treat, such as Douglas fir, are incised with small slits as they pass between spiked rollers to promote penetration (top photo, facing page). Easily treated species, such as Southern yellow pine, are treated without this type of preparation. Generally, lumber up to 1 in. thick is completely penetrated by preservative. For 2-in. thick stock, easily treated woods will be fully penetrated; more-resistant species probably won't. And while the sapwood of some woods is easily penetrated, the heartwood of most resists penetration (photo previous page). Only the sapwood of treated wood has protection against decay greater than what nature already handed the heartwood. Industry standards require 90% penetration of the sapwood thickness of timbers 4 in. thick and up. What's important to remember, though, is that a zone of untreated wood may be exposed during machining, especially in large members.

One indication of the success of treatment is the amount of preservative retained by the wood, which is measured in pounds of preser-

Some woods are harder to treat than others. Small incisions help these to accept preservatives more readily.

The tag attached to the end grain of most treated stock certifies that the wood has been inspected by an independent agency. The agency tag contains information such as the year of treatment, the preservative used and the recommended exposure conditions. This particular piece was certified by the American Wood Preservers Bureau. It has been preserved with chromated copper arsenate (CCA) and is certified for ground contact (0.40 pcf retention level).

vative per cubic foot (pcf) of sapwood. Retention varies with the preservative and the wood, as well as with the wood's intended use. Retention standards are set by the American Wood-Preservers' Association and enforced through chemical analysis of treated wood by an independent third-party agency such as the American Wood Preservers Bureau, the Southern Pine Inspection Bureau, or the Timber Products Inspection Agency. When you buy pressure-treated wood, look for an agency tag or quality mark to ensure that wood has been treated to the retention that's right for its intended use (bottom photo, left). And don't be fooled by lumber stamped "Treated To Refusal." This misleading label is used for wood that is only superficially penetrated.

Of the countless compounds tested as preservatives, only a handful have the safety, effectiveness, permanence and economy that make them commercially important. Preservatives in use today fall into three classes: creosote, oil-borne and water-borne.

Creosote—Creosote is a liquid by-product of the carbonizing of coal into coke. It is highly effective against fungi, insects and marine borers. Creosote injected into crossties, marine piles and bridge timbers in a full-cell process may later bleed into the surroundings, causing contamination. Utility and building poles, freshwater piles, fence posts and industrial woodblock flooring are treated in an empty-cell process that yields a clean, non-bleeding surface. Creosote crossties last about thirty years; utility poles treated with creosote may survive sixty years.

Wood freshly infused with creosote gives off potentially harmful vapors that disappear within a few months. The foliage of plants near freshly treated wood may be killed. Gloves must be worn when handling creosote-treated timbers. Creosote products cannot be painted; coal-tar pitch, urethane, epoxy and shellac are acceptable sealants. Because of its organic origin, creosote eventually biodegrades.

Oil-borne preservatives—Oil-borne products are carried in organic solvents such as liquified isobutane. The preservatives include pentachlorophenol (penta), iodo propynyl butyl carbamate (IPBC), tributyl tinoxide (TBTO), and copper and zinc naphthenate.

Penta will extend a wood's service life by 20 to 40 years. In use since the 1930s, it is used to treat utility and building poles, fence posts and highway timbers. Tinted light to dark brown, penta products glue and finish reasonably well after the noxious oil carrier evaporates. Polyurethane, latex enamel, shellac and varnish are effective sealants.

One problem with penta is that it can migrate to form surface deposits on the wood. It can also leach into the surrounding soil and contaminate the ground water. Penta slowly breaks down into biodegradable compounds.

Until about 1985, exterior millwork was usually dipped in penta carried in light oil. However, IPBC, with its safer solvents, has replaced penta completely in this application. Because of nagging safety and environmental concerns, pundits predict that penta will ultimately fade from the preservation picture.

Water-borne preservatives—The treated lumber and plywood that do-it-yourselfers and builders buy is protected with one of a virtual alphabet soup of preservatives that are carried in water. They include chromated copper arsenate (CCA), ammoniacal copper arsenate (ACA), acid copper chromate (ACC), chromated zinc chloride (CZC), ammoniacal copper zinc arsenate (ACZA) and fluor chrome arsenate phenol (FCAP).

These preservatives share similar chemistries and thus have a lot in common. Chromium holds the other components tightly to the wood and prevents leaching. Zinc and copper fight fungi, while arsenic guards against attack by termites and copper-resistant fungi. Ammonia in ACA and ACZA helps carry copper, arsenic and zinc deeper into penetration-resistant heartwood. Douglas fir and other western woods are commonly treated with ACA and ACZA. Southern yellow pine is usually impregnated with CCA.

CCA's effectiveness, permanence, safety and economy make it the workhorse of the water-borne stable. Virtually every pressure-treated stick sold east of the Missisippi has been treated with CCA, as have the majority of those sold in the west. CCA is even making inroads into the creosote and penta pole markets. Agricultural uses of CCA-treated wood include fencing, plant stakes, arbors and greenhouse flats. CCA products are used mainly in exterior situations where the decay hazard is high, especially where wood is used in ground contact or against concrete and masonry. Above-ground use requires lumber treated to 0.25 pcf retention. Ground-contact use requires 0.40 pcf. Treated-wood foundation systems are treated to a retention of 0.60 pcf (see bottom chart, facing page).

The hallmark of CCA-treated wood is its blue-green tint. It has a residue-free surface and can be used where contact with bare skin is frequent, as in decks and benches. However, watch for the odd piece with a white, gritty surface residue. Though appearing infrequently, this can form when preservative precipitates out of solution during treating in a phenomenon known as "sludging." Leave it with the seller. If on-site, wash the residue off with water, or install the wood so that the residue side will not come in contact with skin.

During the treating process, CCA is water-soluble. After the first day or two of post-treatment airdrying, CCA is rendered insoluble in water in a process called "fixation." During fixation, chromium reacts chemically with the wood, permanently bonding itself, along with the copper and the arsenic, to the cell walls. For this reason, CCA does not leach from wood in service.

Since use began in the 1930s, three basic formulations of CCA have evolved—types A, B and C, which vary by the amounts of chromium, copper and arsenic they contain. The newer type C, or oxide form, is now preferred for most applications. Types A and B have had some surface-blooming problems, while type C has not. Because all CCA is applied as a water solution, no vapors are ever emitted.

No one knows how long CCA-treated wood will really last. Treaters guarantee at least 40 years and consider 100 possible. After being in the ground for 50 years, treated test stakes

Top photo courtesy of Southern Forest Products Association

in Mississippi and Florida still show no decay; untreated control stakes lasted fewer than four years. Surprisingly, CCA-treated wood is not completely immune to insects like carpenter ants, which do not ingest the wood. And because molds live in water, CCA products are susceptible to surface mold and stain if stacked wet, without stickers (solid-piled).

CCA products finish reasonably well. But it is important to remember that when you use pressure-treated wood you're basically using green wood. Because the wood is saturated with water during treating, and rarely kiln-dried afterward, it is often still wet during construction. Water repellents, stains and paints attach themselves to the same sites on the wood cells as does water. Wait at least a week or until the surface is thoroughly dry before finishing. Oil-base semi-transparent stains are the best performers on CCA-treated wood, because they are less affected by swelling and shrinking. If the construction cries for color, follow an oil-base primer with acrylic latex top coats.

At the very least, always use a water repellent that contains mildewcide (oil-base paint can usually be applied later without trouble). Repellents cause rain to bead on and evaporate harmlessly from a wood surface. Without a repellent, repeated rapid swelling and shrinking will carry surface checks into the untreated core of larger members. Renew repellents every couple of years. Water repellents are critical to extending the life of wood, so the latest development in pressure treating is to impregnate wood with preservative and water repellent simultaneously.

Bark side up?—Treated wood demands that you do more than adjust your finishing technique. If you want to get the most out of this product, you must change the way you think about installation. This is especially true in the case of decking, where some of the rules may seem contradictory. Long before treated wood arrived on the scene, for instance, experience showed that exposed decking should be laid bark side up. This was meant to avoid shelling—loosened grain that appears most commonly on the pith side of flatsawn softwood lumber. Shelling occurs when the rapid swelling and shrinking of exterior wood causes the edges of the darker latewood layer of the growth rings to separate from the wood's surface and curl upward. It is common in uneven-grained softwoods, such as Southern yellow pine and Douglas fir.

At least one wood preserver recommends that deckboards be installed bark side up to reduce cupping. But their assumption that decking will cup in the direction of its growth rings is true only if its moisture content at installation is lower than the average outdoor equilibrium moisture content.

What confounds the original wisdom of "bark side up" is that nearly all CCA-treated deckboards are sold water-saturated. After being fastened in place, water-saturated, flatsawn deckboards will cup in the direction opposite the curvature of their growth rings as they dry to the locale's year-round average moisture content for exterior wood. One drawback is that water will puddle on their surfaces. But this really isn't a roblem as long as a water repellent has been applied. A water repellent will also help reduce shelling and its cousin, raised grain.

Correct fastening can help keep cupping to a minimum. For standard 5/4x6 radius-edge decking, a 10d hot-dipped galvanized ring- or spiral-shank nail driven at a slight angle about 1 in. from each edge will do. A smaller nail may pop. Better yet, use the zinc-coated, case-hardened screws developed for this application. From my experience, a good bet is to buy treated lumber a couple of weeks beforehand to permit it to dry partially before building, lay deckboards best-face up and generously apply a water repellent as soon as their surfaces have dried.

Plan for shrinkage when laying pressure-treated deckboards. Many builders gap deckboards with a 16d nail, about 5⁄32 in. But because CCA-treated wood is usually wet during construction, you can end up with bigger gaps than you bargained for. A flatsawn Southern yellow pine deckboard will be about 5⅝ in. wide after treating. Used in New England, it will shrink to about 5 7⁄16 in. as it dries to the region's year-round 16% outdoor average moisture content. A 3⁄16-in. gap will open even if wet deckboards are butted as they're laid.

CCA, ACA and ACC are corrosive to uncoated metal. In above-grade construction use stainless-steel, hot-dipped or hot-tumbled galvanized fasteners. Joist hangers, framing anchors and other hardware should also be corrosion resistant. Types 304 and 316 stainless-steel, Type H silicone bronze, ETP copper and monel fasteners are required for below-grade applications, such as treated-wood foundation systems.

Southern yellow pine is prone to splitting. When nailing or screwing within 2 in. of the end, or close to the edge of lumber, drill a pilot hole. You can reduce splitting by using blunt nails. These punch through wood fibers, rather than cleaving them apart as sharp ones do. Splits caused by careless fastening create water traps that are irresistible to fungi.

Relative heartwood decay resistance of untreated construction woods

Resistant or very resistant	Moderately resistant	Slightly or nonresistant
Baldcypress (old growth)	Baldcypress (new growth)	Pines (other than long-leaf, slash and eastern white)
Cedars	Douglas fir	
White oak	Larch, western	Spruces
Redwood	Pine, eastern white	True firs
	Southern yellow pine: longleaf, slash	
	Tamarack	

Source: U. S. Forest Products Laboratory Wood Handbook

Preservative retention (in pcf*) recommended for timber, lumber and plywood

	Above ground	Ground contact	Permanent wood foundation	Marine
Creosote	8	10	NR	25
Penta	0.4	0.5	NR	NR
Water-bornes				
CCA	0.25	0.4	0.6	2.5
ACA	0.25	0.4	0.6	2.5
ACC	0.25	0.5	NR	NR
CZC	0.45	NR	NR	NR
FCAP	0.25	NR	NR	NR
ACZA	0.25	0.4	0.6	2.5

NR = not recommended
*Pounds per cubic foot
Source: American Wood-Preservers' Association

Where nails are impossible or objectionable, use lag screws or through-bolts. Put washers under bolt and lag heads, as well as under all nuts. The washers will help to distribute stress. Tighten only until snug. Although a common practice, overtightening fasteners to make sure they're "good and tight" is a mistake, because wet or dry, wood under the washer is easily crushed. Most CCA-treated lumber will shrink after it's in place anyway; even properly tightened lag screws and through-bolts will still need to be retightened a few weeks later.

CCA-treated wood glues well. Phenol-resorcinol, resorcinol and melamine-formaldehyde structural adhesives are used in making laminated timbers from treated dimension lumber. On site, use only construction adhesives specifically formulated for treated wood.

Treated wood machines the same as untreated wood. Impregnation with CCA increases abrasion to tools slightly, but you can extend blade sharpness by using carbide-tipped blades. As should be done when machining any wood product, treated or untreated, wear goggles, a dust mask and ear protection.

Remember that larger pieces of treated wood may not be fully penetrated. It's important that you treat site-cut surfaces liberally with an over-the-counter preservative. These preservatives usually contain copper or zinc napthenate, or TBT. Don't overlook pilot holes, mortises, tenons, bevels and site-sawn stair stringers, either. Whenever possible, use precut stringers, railings, post caps and balusters. Always put the "factory end" of treated posts towards the ground. Cap the site-cut surface with flashing or an overhanging railing. Use a continuous railing to avoid the water trap formed when a joint between segments falls atop a post.

Health concerns—Because of the wide use of CCA-treated wood, questions have been raised about its possible effect on health. As a result, these products have come under increasing scrutiny. Most studies done on CCA products, however, have deemed them essentially safe. For example, a 1987 report by the California Department of Health Services stated that "with the possible exception of creosote," none of the common wood preservatives poses a toxic hazard. As with any construction material, however, you should follow certain precautions when using CCA-treated products.

Use treated wood only where such protection is needed. Wear a dust mask and goggles when machining treated wood, and wash your hands after handling it. Wash work clothes separately and before reuse. Do not burn treated wood. Do not use treated wood for cutting boards, countertops, silage or fodder bins, or where it could become a component of animal feed. Though rarely required, CCA-treated wood can be used indoors without being sealed as long as all machining dust is cleaned up.

Only visibly surface-clean CCA-treated wood should be used for playground equipment and picnic tables. For these uses, specify wood treated under AWPA's special standard, C17-88, "Playground Equipment Treated With Inorganic Preservatives." This standard ensures the cleanest surfaces possible. To lessen further the potential for skin contact, apply a water repellent. Or better yet, use an oil-base stain or paint. Round all edges to prevent splintering, and make sure knots and hardware are flush with surfaces.

To help builders use treated wood wisely, the treating industry publishes an EPA-approved Consumer Information Sheet (CIS) for creosote, pentachlorophenol and inorganic arsenical pressure-treated products (it's supposed to be available from treated wood retailers; it's always obtainable from AWPA and AWPB).

If CCA-treated wood is so safe, then why all the precautions? The answer is that where people's health is concerned, it's best to err on the conservative side. An adage among toxicologists says: "The dose makes the poison." Sensitivity to any substance varies dramatically among individuals.

The two routes by which CCA may enter the body are inhalation and ingestion, especially of airborne machining dust. On the one hand a Forest Service employee showed symptoms of arsenic over-exposure after building tables of CCA-treated wood. On the other hand, studies of radial-saw operators and factory employees who build foundation components from treated lumber and plywood found no detectable arsenic concentrations in their systems. The factory workers wore neither masks nor gloves.

There will always be people who are sensitive to the chemicals used in treated wood, as there will always be people who are sensitive to ordinary sawdust. The bottom line appears to be that even without mask or gloves, exposure to arsenic from handling or machining CCA-treated wood is well below the average 80 micrograms Americans get daily from their food and water.

The future—Another water-borne preservative that shows increasing promise is borax, the main ingredient in borates. An old preservative that's been rediscovered, borates protect wood from most fungi and wood-eating insects. Borate-treated wood is unchanged in color from untreated wood. It is noncorrosive to fasteners and can be readily glued and finished. Non-toxic to people and animals, borates also increase wood's fire resistance.

Borates are applied by dipping-diffusion. Green wood is immersed in a hot, aqueous borate bath, then removed and solid-piled. Over a few weeks' time, the preservative naturally dilutes itself by diffusing into the water in the wood. The sapwood is completely penetrated, as is the heartwood of some woods. Dry wood is treated in a full-cell pressure process. Spray application shows promise for treating wood in place. Today, the widest use of borates is in the treatment of timbers for log structures and post-and-beam construction.

The big stumbling block with borates is that they remain water-soluble and thus readily leach out of treated wood that gets wet. Until a way is developed to render them insoluble after treatment, borate products shouldn't be used where they're exposed to weather. Unfortunately, this renders them ineffective in the very environments where treatment is most needed.

Other preservatives on the horizon include chlorothalonil and alkylammonium, or AAC. Chlorothalonil is an oil-borne, EPA registered agricultural fungicide. With performance on a par with penta, only its high cost needs to be overcome. AAC is a water-borne preservative that has already been commercialized in New Zealand.

With the decline in use of creosote and penta, the volume of wood treated with water-borne arsenicals and borates will continue to rise. Simultaneous treatment of wood with preservative and water repellent will grow. Tomorrow's products will combine even lower toxicity levels, with greater effectiveness at low retentions. With good building practice and preservative treatment, wood can last a lifetime. □

Stephen Smulski is an assistant professor of wood science and technology at the University of Massachusetts.

Sources of information

American Wood Preservers Bureau
P. O. Box 5283
Springfield, Va. 22150
(703) 339-6660

American Wood Preservers Institute
1945 Old Gallows Rd.
Vienna, Va. 22182
(703) 893-4005

American Wood-Preservers' Association
P. O. Box 849
Stevensville, Md. 21666
(301) 643-4163

Southern Forest Products Association
Box 52468
New Orleans, La. 70152
(504) 443-4464

American Plywood Association
P. O. Box 11700
Tacoma, Wash. 98411
(206) 565-6600

National Forest Products Association
1250 Connecticut Ave., N. W.
Washington, D. C. 20036
(202) 463-2700

National Institute of Building Sciences
1201 L St., N. W., Suite 400
Washington, D. C. 20005
(202) 289-7800

West Coast Overhang

A framing technique for soffit-covered eaves

by Don Dunkley

Until a few years ago, the typical eave detail on the West Coast was the exposed overhang. When we built ranch-style houses, you could see the rafters and the underside of the roof sheathing (usually shiplapped boards exposed to view). But the current trend where I build in California's Central Valley is toward the Mediterranean look; more and more of the houses we frame have soffit-covered eaves.

In the past, the crew used to grumble and moan when they got to the soffit-framing part of a job. The framework for our first few attempts consisted of sun-baked 2x4s salvaged from the scrap pile. We scabbed them to the rafter tails and toenailed them to a horizontal nailer affixed to the wall studs. Cobbling together row upon row of outriggers around the eave line took a lot of time, and leveling each one was equally frustrating.

About a year ago, however, I learned of a slick method for installing soffits that uses the fascia to support the outboard edge of the soffit. Our crew quickly put the idea to the test, and the job went so smoothly that we now use the method as standard procedure.

In the groove—For soffit material, we typically use ⅝-in. thick plywood with a resawn face, and slip it into a groove cut in the backside of the fascia (top drawing at right). Plywood this thick is stiff enough to span up to about 30 in. without additional support.

Before cutting any rafters, I calculate the position of the level cut and the plumb cut on the rafter tails. First I step off 1 in. on the fascia (usually a 2x10) to allow for a reveal below the soffit. Next I mark off ¾ in. to allow for the soffit groove. I make the groove ⅛ in. wider than the thickness of the soffit material to give me some wiggle room for inserting the edge of the plywood. The remaining width of the fascia represents the maximum depth of the rafter plumb cut. I always lay out the level cut on the rafter to end up a little bit above the groove in the fascia—½ in. to ¾ in. will do. This makes up for the inevitable variations in framing lumber that can bring the bottom of the level cut into the plane of the soffit. I make all the level cuts on the rafters before stacking the roof.

The top edge of the fascia should be beveled to match the pitch of the roof. Most of the lumberyards that I deal with can provide this service. If yours doesn't, take the time to bevel the fascia yourself. It will give you a consistent line at the roof edge and solid bearing for the roof sheathing.

I use my router, guided by a 1x4 tacked to the back of the fascia, to plow the groove. My Skil model 5000 router is closing in on 20 years old, and it's rated at a mere ⅞ hp. But with a ¾-in. carbide bit it still gets the job done (though it takes two passes to get to the full ½-in. depth of the groove).

I rip my soffit plywood about ¼ in. shy of the overall dimension between the stud wall and the bottom of the groove in the fascia. This makes it easy to slip the plywood into the groove without the free edge hanging up on slight bows along the wall.

To vent the roof we drill trios of 3 in. holes, 3 in. apart. The center holes in each group are 4 ft. o. c. (bottom drawing at right). Our standard practice is to start the center hole in the first trio 10 in. from the beginning of the sheet. Drilled on this layout, the spacing between the holes is consistent from sheet to sheet. To keep the bugs out, we staple insect screen over the holes on the concealed side of the plywood.

The inboard edges of the soffits are affixed to nailers with hand-driven 8d hot-dipped galvanized nails. I locate the bottom edge of the nailer by first leveling across from the bottom of the fascia to the wall. I mark this point, and measure up from it the distance from the bottom of the fascia to the top of the groove. Measuring down from the top plate of the wall to this mark gives me a constant reference point for positioning the nailer. It's a good idea to put building paper on this part of the wall before installing the nailer. (If the wall depends on plywood for shear resistance, it must go on before the nailer.) I let the paper hang down below the nailer far enough so that the siding or stucco crew can tuck their paper under it. If there are openings above the nailer between the stud bays and the soffit, I draft stop them with pieces of plywood.

With this system down pat, my crew can run a substantial amount of soffit in a day's work. The only limiting factor seems to be the time involved in setting up the scaffolding for the predominantly two-story houses that we build. □

Don Dunkley is a framing contractor living in Cool, California.

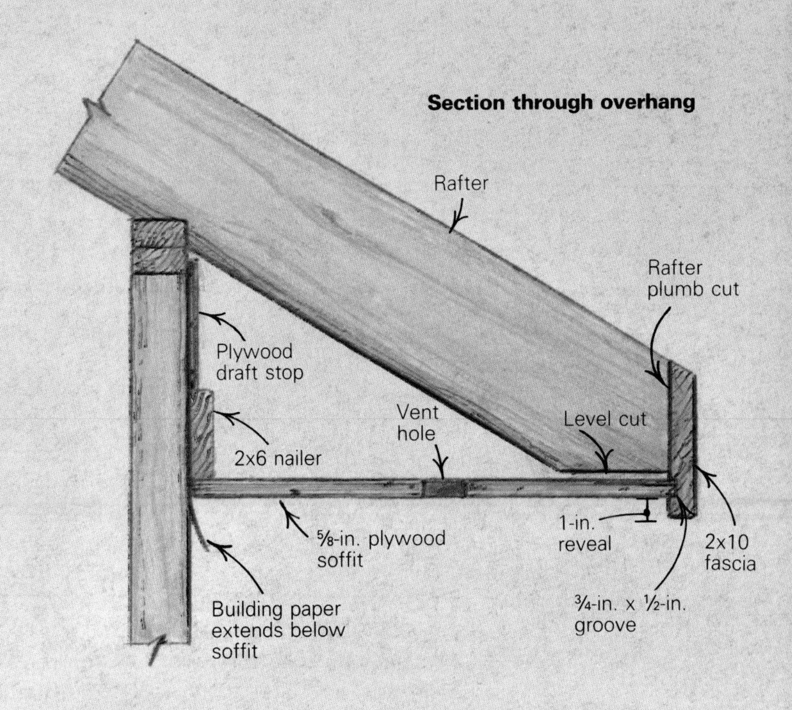

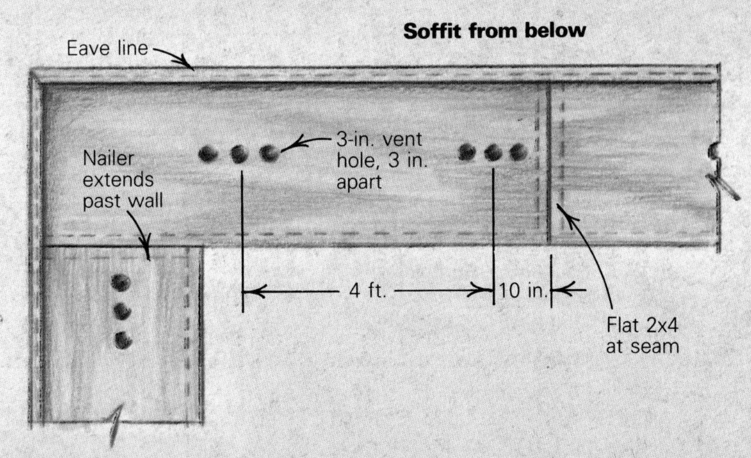

Drawings: Bob Goodfellow

From *Fine Homebuilding* magazine (October 1990) 63:69

Falling Eaves

One carpenter tries five methods of framing a tricky roof

When an angled bay is capped by an extension of the main roof, the rafters that sit on the angled walls require some careful cuts.

by Scott McBride

Roof framing is tricky enough when the walls are plumb, level and square, but when the rafters intersect an eaves wall that's angled, well, it'll drive a person to thinking. The first time I encountered this condition was with a 45° angle bay extending from a single-story exterior wall (photo above). Instead of having its own separate hip roof, broken up into the usual three planes, the roof of this bay was simply to be an extension of the main roof plane, intersecting the diagonal walls of the bay on an angle in both plan and elevation.

The cornice in this situation runs at some oblique angle in plan (usually 45°) while it falls in elevation. I call this condition "falling eaves," as opposed to regular eaves, which run level.

I improvised my way through that job, thankful that only a few rafters were affected. But when the time came on another house to frame a gable roof with all four of its corners lopped off in this fashion, I decided to study the problem carefully. I ended up with five diferent solutions.

An octagonal room—I was framing a new house that featured an octagonal room extending above the main roof (photo left, facing page). Unlike most octagonal roofs, which have eight roof planes coming to a point at the peak, this roof was essentially a gable, with only two roof planes meeting along a ridge. Falling eaves were located where the angled walls of the octagon intersected this roof.

I began by framing the two gable walls and the two regular eaves walls. To illustrate my method for calculating the heights of these walls, I'll simplify the dimensions a bit. Let's say the level eaves walls were 8 ft. high, and the run of the roof was also 8 ft. (half of a 16-ft. span). Angled walls chop off the four corners of the room (drawing, p. 122), extending in 4 ft. from what would have been a square corner. In that 4 ft. of horizontal run, the 9-in-12 roof rises 4 in. x 9 in., or 36 in. total. The height of the gable wall at its outside corners (the lowest points) would therefore measure 8 ft. plus 36 in., or 11 ft. total. Actually, there's an adjustment that I had to make here, which I'll discuss momentarily.

From its outside corners, the gable wall rises 36 in. over the 4 ft. of run. That would make the height of the gable wall at the peak 11 ft. plus 36 in., or 14 ft. total. But there was a further complication in calculating the height of the gable walls. The calculations just given start at the outside corner of the eaves plate. This point lies on the measuring line, which runs somewhere down the middle of the common rafter. But the gable walls needed to support lookouts for a framed rake overhang (photo top right, facing page). The lower edges of these lookouts line up with the lower edges of the common rafters, in a plane below the measuring line. Consequently, the gable-wall top plates had to be lowered by an amount equal to the vertical depth (heel cut) of the common-rafter bird's mouth.

Falling headers—After tipping up the level eaves walls and the gable walls, I connected them with four sloping headers, which I'll call falling headers. Later, I added 2x6 studs under these headers to frame the angled walls. As with the angled walls of the bay described earlier, the top edges of these falling headers are similar to the top edge of a valley rafter. Because the header travels a horizontal distance of 4 ft. perpendicular to the level eaves wall, I knew that its diagonal run in plan would be 17 in. multiplied by four. The hypotenuse on a right triangle with 12-in. sides is 17 in. (actually it's 16.97). The rise in each of those 17-in. diagonal increments of run had to be the same as for each 12 in. of common-rafter run (namely, 9 in.) in order to keep the header aligned with the roof. To find the actual length of the falling header on its long face, therefore, I stepped off 9-in-17 with the square four times. (A shortcut would have been to multiply by four the number listed under 9 in the length-of-hip/valley rafter table on the framing square.)

Finding my calculated length precisely consistent with the field-measured distance from level eaves wall to gable wall (well, close enough), I drew parallel plumb cuts on the header's outside face at both ends. Through these lines I made opposing cheek cuts, with the blade of the circular saw tilted at 45°. I made four of these headers and spiked two of them into their adjoining walls with 16d nails (I had different plans for the other two headers).

Photo above: Kevin Ireton

Lopping off the corners. **This octagonal room is sheltered by a gable roof that has its four corners lopped off. To find the best way to frame the intersection of the rafters and the sloping, angled eaves walls, the author framed each of the four corners a different way. Photo by Kevin Ireton.**

Gable overhang. **The gable walls were shortened an extra few inches to allow for 2x10 lookouts, which cantilever beyond the walls to form the overhang. This framing assembly is called a ladder.**

It's not supposed to line up. **At its lower end, the header in the photo below aligns with the eaves wall. Because the gable wall was lowered to allow for the lookouts, the header protrudes above the gable-wall plate to run parallel with the roof.**

Method #1: raised block. **Nailing triangular blocks on the sloping, angled headers created a level surface to seat the rafters on. But this required a deep heel cut on the bird's mouth and weakened the rafter overhang.**

Because the outside corners of the falling headers were aligned with the level eaves plates at their lower ends, they could not align with the gable-wall top plates at their upper ends and remain parallel to the roof surface. This is because the gable-wall top plates were recessed below the roof surface by the full vertical depth of the rafters (to make room for the lookouts). However, the eaves-wall plates were recessed below the roof surface by the raising distance (the vertical depth of the rafter above the plate). The raising distance takes up only part of the vertical depth of the rafter, with the heel cut of the bird's mouth taking up the remainder (drawing detail next page). Therefore, the top ends of the falling headers protrude above the gable-wall top plates (middle photo, right). The height of the protrusion is a vertical distance equal to the heel cut of the common-rafter bird's mouth.

I had several ideas about how I might frame rafters into these falling headers. To find out which was best, I decided to frame each of the four corners of the room in a different way.

The raised-block method—Because I'm used to seating the lower ends of rafters onto a level surface, I first tried building up level bearing on one of the falling headers by adding triangular blocks (bottom photo, right). To lay out the spacing for the jack rafters, I pulled 16-in. centers from

From *Fine Homebuilding* magazine (April 1992) 74:82-85

Method #2: the notch. **Another way to create level bearing for the rafters was to cut a notch in the header with a handsaw. Although it worked, this method did not provide much bearing surface for the rafters.**

Method #3: beveled header/level seat cut. **Here the falling header is beveled, as is the seat cut of the rafter's bird's mouth. The intersection between the rafter and the header is a level line (above left). The bevel angle of the header was determined by the top of the eaves-wall plate where the two intersect (above right). The author scribed a scrap block in place to determine the angle. This method provided good bearing and was the easiest to nail.**

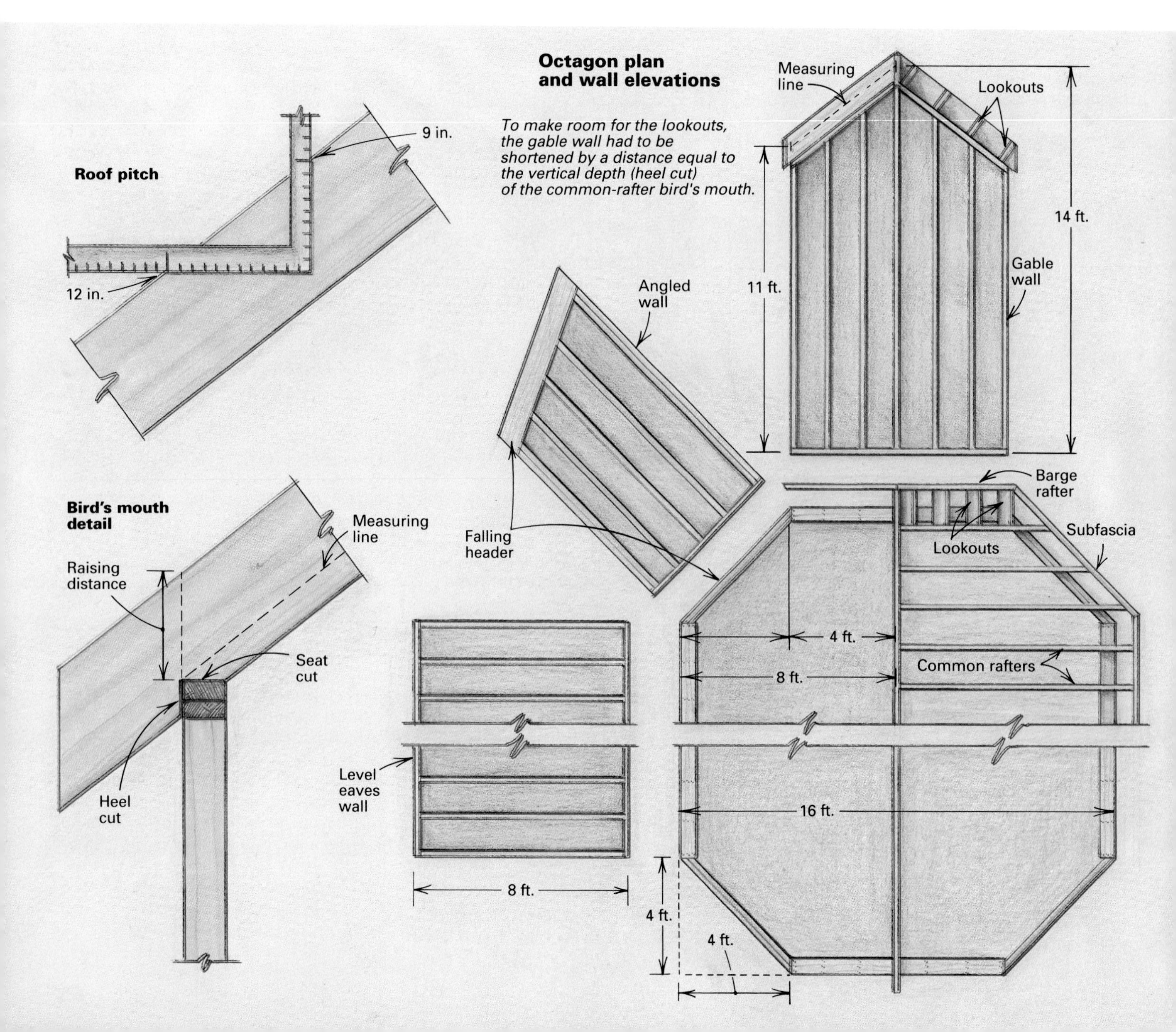

Method #4: beveled header/no bird's mouth. **Here the header is beveled in plane with the bottom of the rafters. There is no need for bird's mouths; the rafters simply bear on the headers (above left) and are held in place by toenails. The bevel angle for this header was determined by the top of the gable-wall plate where the two intersect (above right). The bottom end of the header is recessed below the level eaves-wall plate.**

Method #5? **From this view, you can't tell whether the header is notched for the rafter or vice versa. The author tried the former but thinks the latter method, which he tested on a model, might be the best.**

the nearest common rafter, first making sure the common was straight. The blocks themselves were laid out using the numbers 9 and 17 on the rafter square. With the top of the block presenting a level surface, I could put an ordinary 9-in-12 square seat cut on the jack rafter (cutting on 12). The heel cut of the bird's mouth for these rafters, as well as for the rafters on the other three corners, was made by laying out the standard 9-in-12 plumb cut (cut on 9) on the face of the rafter but cutting it with the circular saw tilted 45°.

As you can see in the photo, the bird's mouths for the blocked-up rafters had to be cut quite deep. This was necessary because the triangular blocks protruded above the top edge of the falling header, which reduced the raising distance and increased the vertical depth of the bird's mouth. Structurally, this weakened the overhang.

The notch method—Another way of producing a level surface on which to seat the rafters was by notching the header. First I reasoned that any line drawn square across the edge of the unbacked header would be a level line. After squaring such a line across the header's edge, I extended a level line on the inside face of the header from the point where the squared-across line hit the header's inside face (photo left, facing page). Along this line, I measured off the 45° thickness of a double 2x—about 4¼ in. From that terminus, I plumbed up to the top inside edge of the header. From where the plumb line struck the corner, I connected back to where I started from on the outside face of the header. These three lines described the two handsaw cuts—one plumb, the other level—that I needed to make the notch.

The bird's mouth to fit these notches was essentially the same one used for the raised-block method, except that it didn't need to be cut extra deep. Because the outside corner of the notch lines up with the outside corner of the level eaves plate, I used the standard raising distance.

Although leaving the strength of the rafter uncompromised, the notch method offers a small bearing surface, which could be a problem with long or heavily loaded rafters.

Beveled header/level seat cut—On one side of the octagon, I used a header backed (meaning that its top edge was beveled) so that its intersection with the seat cut of the rafter was a level line (middle photo, facing page). To find the correct backing bevel, I took a 2x scrap and put a 9-in-17 plumb cut on it sawn at a 45° blade tilt. The scrap mimicked the cheek cut on the end of the header. Holding the cheek cut of the scrap vertically against the end of the level eaves-wall top plate, I scribed a level line across the plate onto the end grain of the scrap. This showed me how much to take off the inside edge of the header (right photo, facing page). Using a wormdrive saw equipped with a rip fence, I beveled one of the headers that I hadn't already installed and then spiked it in place.

To fit the rafters to this header, I laid out the standard common-rafter bird's mouth and made the heel cut (plumb cut) of the bird's mouth with the saw blade tilted at 45°. The seat cut was also made with the blade tilted, but not at 45°. For this I had to use the plumb-cut angle of the common rafter. This method provided good bearing and was the easiest to nail. The backing operation, however, took some time.

Beveled header/no bird's mouth—To round out my experiment, I backed the last header so that its top edge would lie parallel to the roof plane. The big advantage of this method was that the rafters required no bird's mouth whatsoever, so there was no need to calculate precise rafter length (I cut the rafter tails in place later). I was able just to lay the rafter stock down on the mark and toenail (photo above left). I once saw ordinary rafters framed the same way, sitting on walls with tilted top plates, but I suspect that this connection might slip over time (if used with ordinary stud walls) because of the thrust of the roof. In this case, however, I felt that plenty of spikes driven into a beefy header (doubled 2x10s) would adequately resist the lateral load.

To determine the header's bevel, I used the same scrap-block trick I had used for the preceding method, except that I scribed it against the gable-wall top plate rather than against the level eaves-wall top plate (middle photo, above). Unlike the other three falling headers, which are aligned with the level eaves plate and protrude above the gable-wall plate, this header aligns with the gable-wall plate at the top and is recessed below the level plate at the bottom. It's offset from the roof surface by the rafter's full depth rather than by just the raising distance.

And the winnner is—You may be wondering which method I like the best. Well, the beveled-header/no-bird's-mouth method was the easiest. But given my concerns about the rafter-to-plate connection slipping over time, the beveled-header/level-seat-cut method is probably the best of the four I tried.

Since completing the project, however, I have thought of another method. As I looked at a photo of rafters installed using the header-notch method (photo above right), I realized it's impossible to tell from the uphill side whether the rafters are let into the header or vice versa. Instead of notching the header, I could have made a sloping beveled seat cut on the jack rafter's bird's mouth that would mate directly with the edge of the unbacked header. I developed the angles for this method on paper and tested them on a scale model. It works.

Then there was that guy from California who told me he just uses metal framing anchors... □

Scott McBride is a contributing editor of Fine Homebuilding *and lives in Sperryville, Va. Photos by author except where noted.*

Drawing: Bob Goodfellow

Framing With The Plumber in Mind

A few tips to help you keep your sticks and nails out of my way

by Peter Hemp

Contrary to what you might believe, most plumbers, including yours truly, do not enjoy chopping a house to pieces in order to get plumbing systems in place. However, I do claim ownership of a carborundum-tipped chainsaw for just this purpose.

When cost estimating, whether I'm on a job site or working from a set of blueprints, I first start looking for the amount of chop time that's necessary to get the rough plumbing in place. The more wood that I can leave untouched, the cheaper my labor bill is going to be.

Many things will directly influence the labor cost of installing the rough plumbing, including the location of windows, medicine cabinets, let-in bracing, beams and the HVAC ductwork. But some of the most important factors are the direction and position of the floor joists and the availability of unobstructed pathways for vents.

Even minimal plan changes might need clearance from a higher authority, so I'd like to clue you in to those areas you can improve upon with little fuss, very little cost and all by yourself.

Laying down the joists—In the past, I have severed innumerable floor joists because they interfered with toilet wastes or tub and sink drains. Having done some time on framing crews myself, I remember how easy it was to lay joists from one end of the house to the other without thinking about the location of the walls until the subfloor was down. But in relation to plumbing costs, it's here that the greatest savings in labor (mine and yours) and the biggest gain in structural integrity can be realized.

On your next set of prints, look at the location of the bathroom wall behind the toilet. I call it the "tank wall." For standard toilets, the measurement from the finished wall surface to the center of the toilet drain (or closet bend) in the floor is usually 12 in. I like to use 13 in., which gives more clearance behind the tank for future paint jobs (drawing below). I'll need a minimum 3 in. of clearance around this center point to rough in ABS plumbing (about 3½ in. for no-hub cast-iron pipe). This allows room for the pipe (typically 4½ in. in diameter), plus the added hub diameter of the closet flange (for more on toilet installation, see *FHB* #46, pp. 54-57).

If there's a floor joist in that forbidden territory, I'm afraid it's recoil-starter time. When you plan to run joists 16 in. o. c. starting from the tank wall, center your first one about 8½ in. from the finished wall. You can then add a joist on the opposite side of the drain, again placing it at least 3 in. from the center, depending on the type of pipe used. Place an additional joist

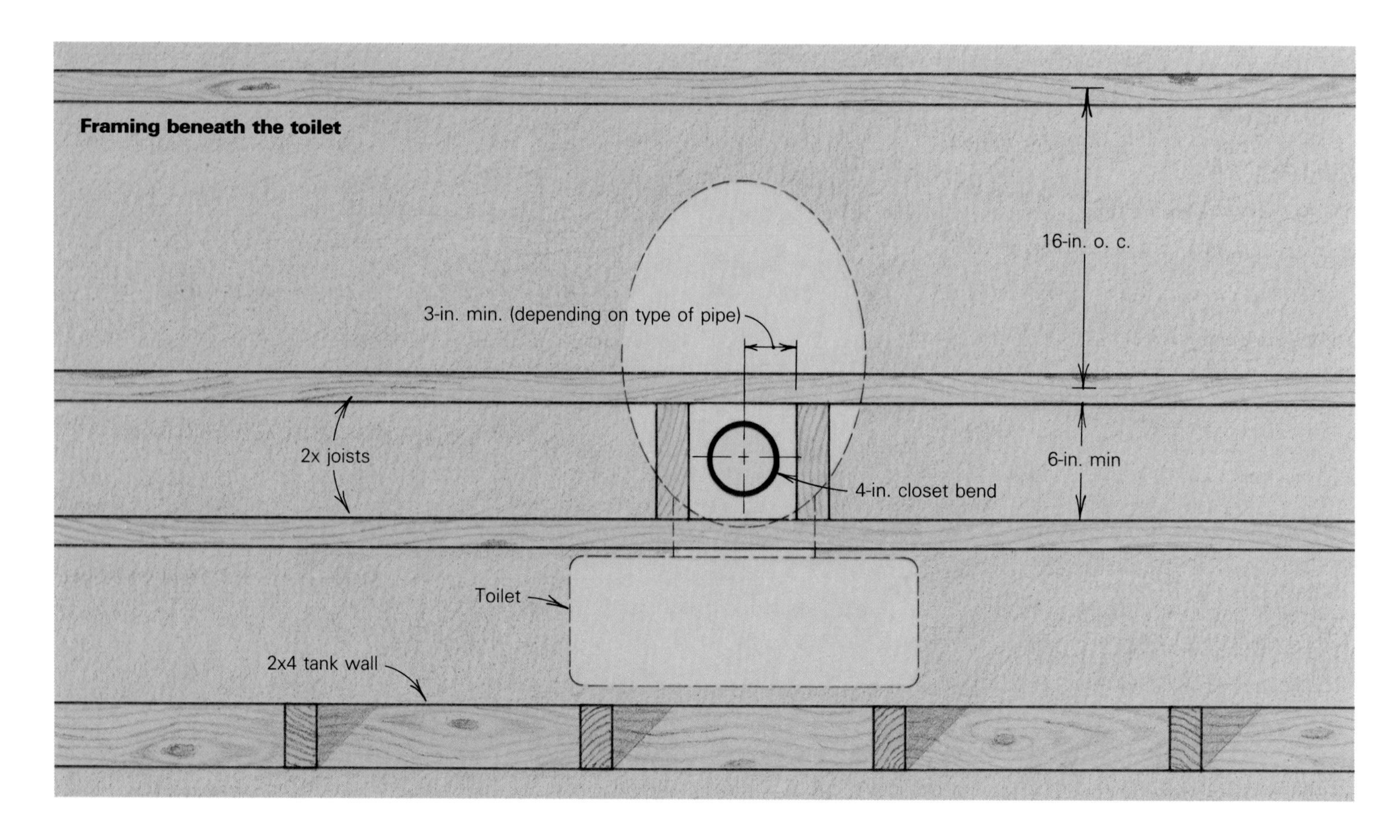

Drawings: Vince Babak

wherever it's needed to satisfy the floor's structural needs.

If joists run perpendicular to the tank wall, I'll still need at least 3 in. of clearance around the center point of the closet flange. If there's a joist in the way, shift it over.

If you want to do a good deed for your customer after I've installed my drain, waste and vent system, and it has passed inspection, add some blocking around the closet drop. This blocking will stiffen the floor under the toilet, and that will help to prevent seepage from the toilet for the life of the structure. It will also save the owner lots of money later for the replacement of the subfloor and floor coverings.

Don't forget to take bathtubs into account when laying out the joists. Where joists run parallel to the length of the tub, I'd like a joist on either side of the drain hole, 6 in. from the center, with a block about 12 in. away from the shower-head (or valve) wall to support the pipe (drawing below). This distance will vary according to the type of tub and drain pipe, so check your tub before nailing up the block. When joists run perpendicular to the length of the tub, don't put a joist any closer than 12 in. to the valve wall if you can help it. If the toilet and shower head flank the same wall, you can't satisfy the requirements for both. In that case, leave a joist 8 ½ in. from the tank wall. I'll notch the top of the joist for the tub shoe (the drain fitting) if it's in the way and drill through the joist for the tub drain.

Again, you may have to add one or more joists to maintain proper joist spacing beneath the tub. But when you compare the cost of a couple extra joists to the cost of replacing a severed joist in not-so-perfect working conditions (how would *you* like crawling around in the dark over moist, unidentifiable objects?) you'll find the cost of the extra joists a real bargain.

Framing for tubs—While we're on the subject of tubs, most residential tubs are between 30 and 32 in. wide. When carpenters frame the valve wall, they often put a stud about 16 in. from the adjoining wall. What a shame. How many times have you sat or stood in a tub and visually lined up the strainer in the bottom of the tub with the waste and overflow plate above it, the tub spout, the tub and shower valve and finally the shower arm and head? Did it bother you to see them all out of whack? Well, this all goes back to putting that stud in the middle of the valve wall. I have to carve up the stud to anchor my valve, spout and shower arm, and it never comes off looking professional.

Instead of that one stud in the center, please install two, dividing the width of the tub into three equal spaces. That way, I can add blocking between the two studs to anchor my plumbing fixtures. The blocking also makes it easy to line up the fixtures vertically.

Troubles at the perimeter—My final gripe about floor joists has to do with those lying beneath exterior 2x4 plumbing walls. When the joists run perpendicular to the plumbing wall, I have to contend with either a 2x rim joist or a rim joist with a row or two of 2x blocking. When the joists run parallel to the wall, I'll usually find a single joist toenailed to the mudsill.

When I have to bring 2-in. drain pipes (2½-in. O. D.) down through the bottom plate of the wall, the structure suffers. For the lines to stay completely inside the wall, (drywallers love you when they don't) I have to run a 2½-in. pipe through, at best, about a 2-in. space. If there's a fitting on the drain within the height of the joist, which is often the case, I need an additional ¼ in. to ⅜ in. of clearance. What does this mean? It means I have to remove ½ to ⅞ in. from the inside of the rim joist or block (if there's an inner block, I have to tear it out first). Worse, my drill makes a 2⅝-in. hole when it's sharp. When it's dull, it travels out-of-round and adds another ⅛ in. to that. So the joist or block can wind up being just ½ in. to 1 in. thick, give or take a few hairs. That's if my vertical cutting, done with the long rough-in blade in my reciprocating saw, is perfectly plumb, which it usually isn't. And that's before I chisel out between saw cuts.

You might be wondering why I don't use smaller pipes. If the same plumber who installed new pipes in a structure had to come back later to unclog them, there wouldn't be any drain lines smaller than 2 in. in a house. This happens to be my credo. Though our local building code calls for 1¼-in. drains for lavatories, 1½-in. drains for tubs and 2-in. drains for kitchens and laundries, I prefer to use 2-in. drains for all of them. If I do use a smaller diameter drain in a 2x4 wall, say 1½ in., it still means chewing away ¼ to ½-in of wood. I could reduce the diameter of the pipe by using DWV copper (which has half the wall thickness of supply-line copper) instead of ABS plastic or no-hub iron pipe, but

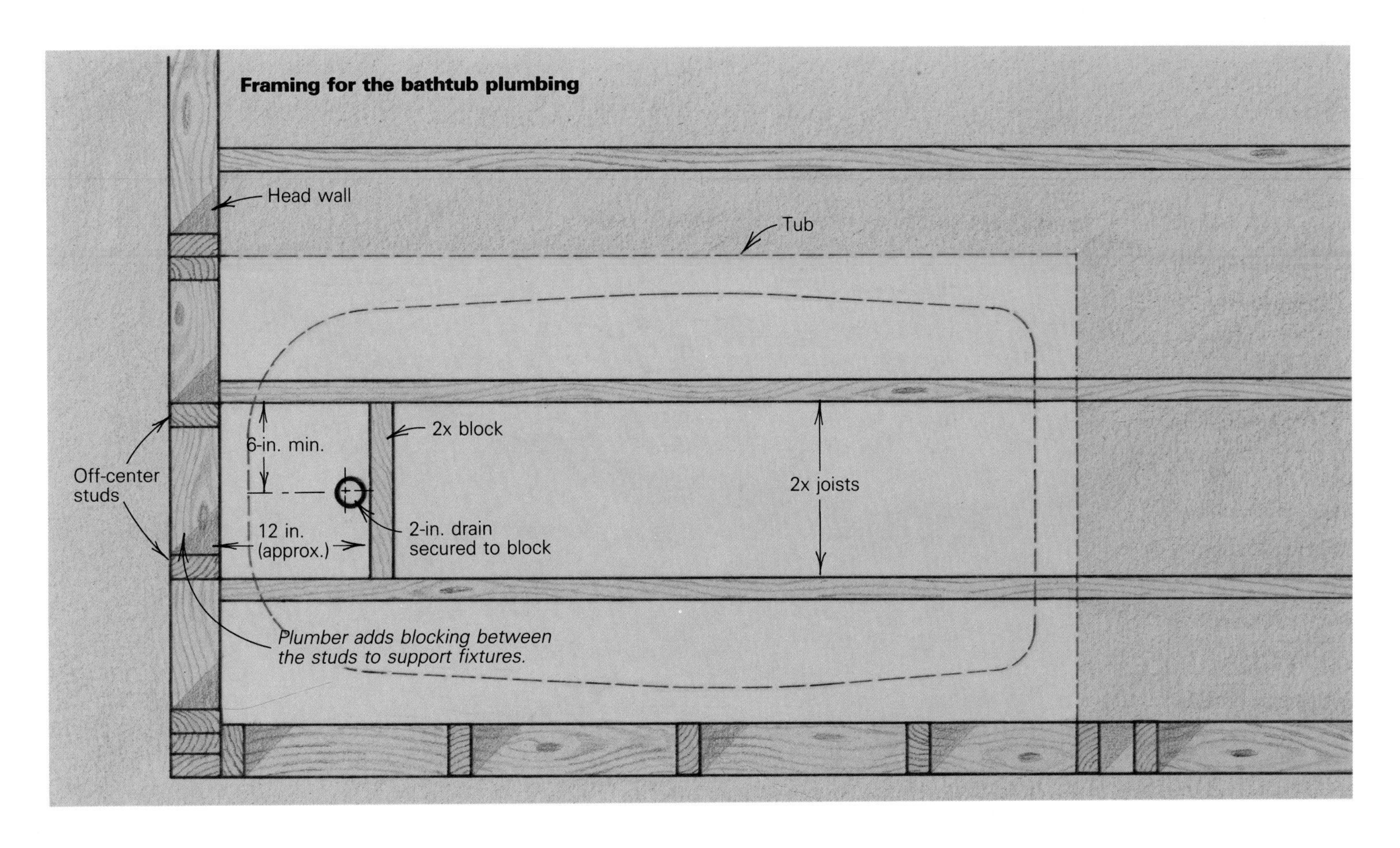

the cost of the copper pipe and fittings and of soldering the joints is considerable. Also, the torch I need for soldering could start a fire (if you don't think that plumbers are guilty of starting a few fires, think again).

So what is the remedy for all this? It's simple: furred walls. Nail 2x2s to the studs, and that extra 1⅝ in. will help me to stay clear of the rim joist or end blocks. You'll spend a few extra bucks for materials, but I'll bet in labor (yours and mine), furring the wall is cheaper than paying me to crawl and hack. And the structure of the house suffers less. If the rim joists parallel to the mudsill are doubled, which is sometimes the case, I'll still have to chew through the inner joist, but the furring will allow me to keep the outer joist intact. Of course, 2x6 walls would be even better than furring, or the designer can design the house to keep plumbing out of exterior walls in the first place.

Nails in the plumbing zone—A plumber spends most of his time boring holes. Unlike those "sparkies" who rarely have to drill any hole larger than a finger for their skinny little wires, we plumbers occasionally find ourselves boring 5-in. holes. And, the worst thing to encounter when boring big holes is a nail.

For fast drilling in wood, plumbers like to use self-feeding bits, which means we try to hang on to that powerful Milwaukee Hole Hawg, waiting to have it wrenched from our hands, pin our wrists, slap us in the face with the handle or spin us right off a ladder while the bit chews its own way through your masterwork. These drill bits can easily cost $25 or more and can be destroyed by nails the first day out of the box.

When I encounter a nail, I stop drilling immediately (as a safety feature, there is no trigger hold-down button on a Hawg; to have one would be suicide). I then have to remove the self-feed bit, replace it with a hole saw and continue drilling at a much slower rate, expending back-breaking energy. Now, if you were merely to stab your 16d sinkers a little differently, you would save me a lot of hassle and yourself some money.

Here's the program. Before you nail up a wall, consult your funny papers (some call 'em blueprints) to locate the plumbing fixtures. Next, mark the stud bays where the pipe is supposed to go, plus one bay on either side. Then, when you're nailing the bottom plates to the subfloor, start the nails in the middle of the marked bays, but don't sink them entirely. That way, if they interfere with my boring path, I can yank them out, move them over and set them myself. This also gives me the option of running the pipe through the adjacent bays if I have to. Do the same thing with the top plates; I'll sink all the nails when I'm through.

If the wall needs blocking, set the nails in the marked bays just enough to hold the blocks in place. I can bore the lower and upper plates, yank the blocks and bore them separately. Then I can slide them over my pipe (if there aren't any couplings in the way) and nail them in place myself. If the holes need to be bored near the ends of the blocks, the nails will usually split them. To avoid this, I use my cordless screw gun to predrill the blocks and screw them in place.

Backing the drywall—If you aren't asleep yet, hang on for the finale. My final suggestion deals with backing for the ceiling drywall. It used to be that a 1x6 was used for backing on top of a 2x4 wall, and a 1x8 backed a 2x6 wall. These days, two ceiling joists are often used instead, each slightly overlapping the opposite edges of the top plate (drawing below). This creates a U-channel, which can make it very tough for me to run my pipes.

If the plumbing wall is anywhere near the bottom slope of the roof, and the roof sheathing is already down, working in that location is almost impossible. The U-channel also makes it tough to tighten no-hub couplings, which are most often down in the channel. And there's more. Sometimes I try to avoid extra vents in the roof by back-venting several 2-in. vents into one 4-in. stack. The U-channels can end up getting so mauled that I have doubts as to their integrity.

On interior plumbing walls, I'd recommend using the 1x backing and avoiding center-span nailing in the plumbing bays. As I did for the plates and blocks, I'll finish the nailing for you after my pipe is through.

It may not be evident at first, but I think you'll find that my recommendations won't just save me labor, they'll save you some painful rehab activity that you would just as soon not experience. □

Peter Hemp is a plumber and writer from Albany, Calif., and author of The Straight Poop *(Ten Speed Press, P. O. Box 7123, Berkeley, Calif. 94707, 1986. $9.95, softcover; 176 pp.) a book on plumbing maintenance and repair.*

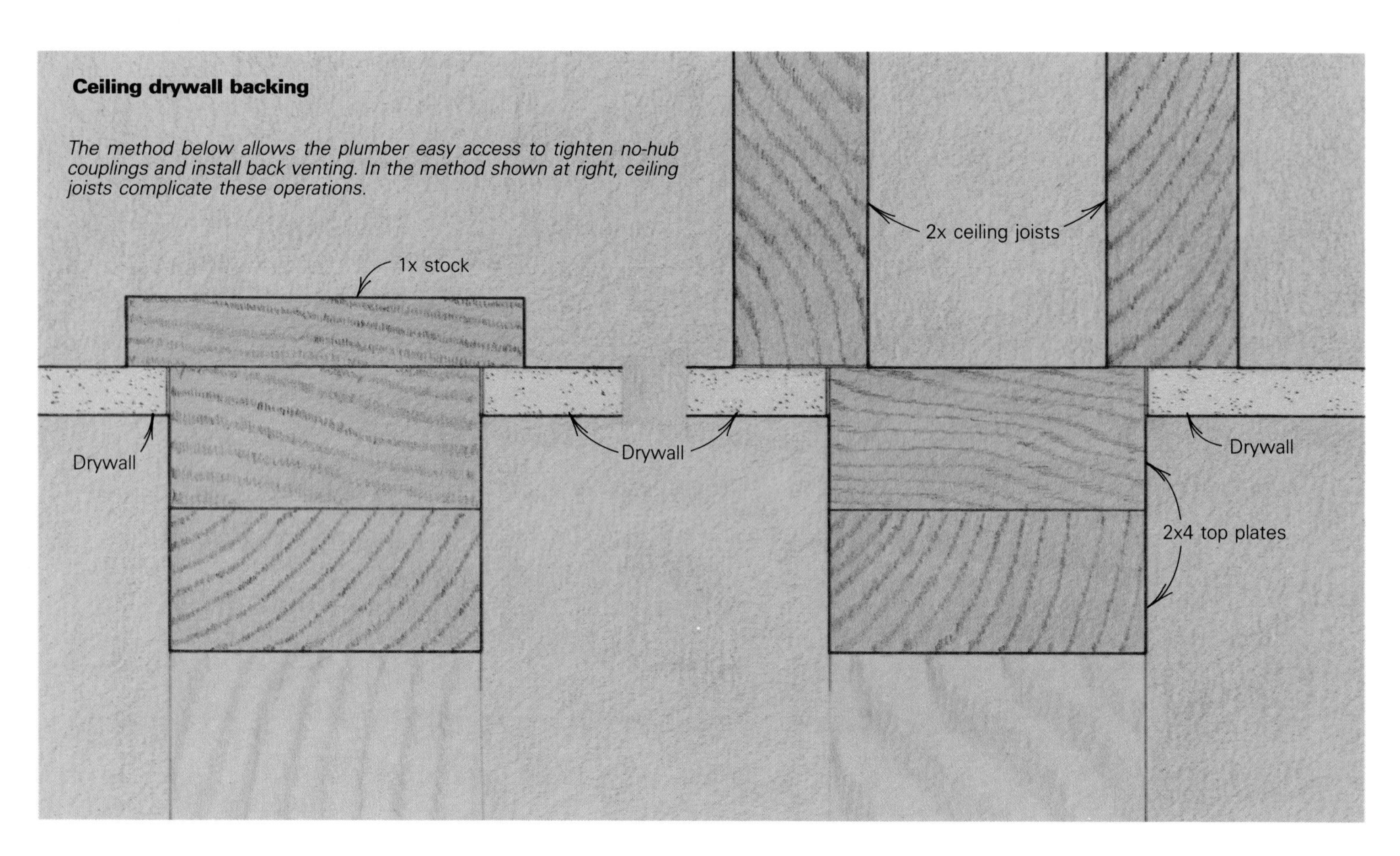

INDEX